RELIABILITY
ENGINEERING

PROCEEDINGS OF THE ARAB SCHOOL ON SCIENCE AND TECHNOLOGY

R. Descout (editor)
**Applied Arabic Linguistics
and Information & Signal Processing**

A. H. El-Abiad (editor)
Power Systems Analysis and Planning

T. Kailath (editor)
Modern Signal Processing

E. E. Pickett (editor)
Atmospheric Pollution

G. Warfield (editor)
Solar Electric Systems

P. D. T. O'Connor (editor)
Reliability Engineering

FORTHCOMING

R. Risebrough (editor)
Pollution and the Protection of Water Quality

P. A. Mackay (editor)
Computers and the Arabic Language

RELIABILITY ENGINEERING

Edited by

P. D. T. O'CONNOR
British Aerospace

⬤HEMISPHERE PUBLISHING CORPORATION
A subsidiary of Harper & Row, Publishers, Inc.

Washington   New York   London

DISTRIBUTION OUTSIDE NORTH AMERICA

SPRINGER–VERLAG

Berlin   Heidelberg   New York   London   Paris   Tokyo

RELIABILITY ENGINEERING

1 2 3 4 5 6 7 8 9 0   B C B C   8 9 8 7

Library of Congress Cataloging-in-Publication Data

Reliability engineering

(Proceedings of the Arab School on Science and Technology)

''Proceedings of the seminar on reliability held by the Arab School on Science and Technology in Zabadani, Syria, during July 1986''—Fwd.

Bibliography: p.

Includes index.

1. Reliability (Engineering)—Congresses.
I. O'Connor, Patrick D. T.   II. Arab School on Science and Technology.   III. Series.

TA169.R439   1988   620'.0042   87-17789

ISBN 0-89116-684-X   Hemisphere Publishing Corporation

DISTRIBUTION OUTSIDE NORTH AMERICA:
ISBN 3-540-18299-3   Springer-Verlag   Berlin

Contents

Contributors

Patrick D. T. O'Connor
British Aerospace
Army Weapons Division
PO Box 19
Six Hills Way
Stevenage
Hertfordshire SG1 2DA

Robert R. Britton
Plessey Assessment Services Ltd.
Abbey Works
Titchfield
Fareham
Hampshire PO14 4QA

Dr. R. N. Allan
University of Manchester Institute
of Science and Technology
Department of Electrical Engineering
and Electronics
PO Box 88
Sackville Street
Manchester M60 1QD

Dr. Nihal Sinnadurai
British Telecom Research Laboratories
Martlesham Heath
Suffolk

David Hutchins
DHA Ltd.
13/14 Hermitage Parade
High Street
Ascot
Berkshire SL5 7HE

David Newton
Department of Engineering Production
University of Birmingham
Edgbaston
Birmingham B15 2TT

Salim Hariri
Electronic Department
Electrical-Mechanical Engineering
College
Damascus University
Syria

Abdullah Shaalan
and Nasser Al-Mohawes
Electrical Engineering Department
College of Engineering
King Saud University
Riyadh, Saudi Arabia

Preface

This book is an edited version of the proceedings of the seminar on reliability held by the Arab School of Science and Technology in Zabadani, Syria, during July 1986. The seminar was aimed at presenting delegates with information on recent developments in reliability engineering and management related to the principal needs of the Arab countries. A team of speakers, all from the UK, presented the seminar. Two specialist contributions were made by Arab authors. I was privileged to have been invited by the Arab School of Science and Technology to lead the seminar and to provide some of the contributions.

The seminar was directed primarily toward the presentation of modern practical engineering and management approaches to attaining high reliability of present technology products and systems. This reflects the move away from the rather academic approach to reliability presented by many courses and texts in the past. The seminar objectives specifically excluded detailed teaching of basic techniques in reliability. Therefore, only a brief resume of basic mathematical and probability techniques is given, but references are provided to texts that cover such material in more detail. Also, the book is not a comprehensive treatment of reliability engineering and management, but it is hoped that readers will find useful material on the new methods and technologies covered.

Since production quality control is such a major determinant of the reliability of modern products and systems, considerable coverage is given to recent developments in this area, particularly in relation to quality circles. Many, if not most, modern products and systems now rely heavily on digital microelectronic systems, so microelectronic devices and software are also covered in some detail. The other major topics covered also reflect recent developments, as appropriate.

Part of the book is extracted from "Practical Reliability Engineering" (2nd edition), written by myself and published by John Wiley and Sons. I am very grateful for their permission to use this material.

Finally, I would like to thank the staff of the Arab School of Science and Technology for the opportunity to participate in the seminar. It has been a most enjoyable and stimulating experience, for myself and for the other contributors.

P. D. T. O'Connor

ix

The Arab School on Science and Technology

HISTORY

The Arab School on Science and Technology, a pan-Arab, nonprofit organization headquartered in Damascus, Syria, was founded in 1978 by the initiative of the Kuwait Institute for Scientific Research (KISR) and the Scientific Studies and Research Centre (SSRC) in Syria to provide a high level continuing education program to Arab scientists in fields that are judged crucial to the development of the Arab countries.

Since its establishment the School has dealt with four major topics: electronics, energy, environment, and informatics. The School has plans to expand into other areas of specialization. A list of the School's previous and forthcoming sessions is given below.

The School has attempted to create for each topic a regular forum of scientific exchange in chosen areas of the Arab region. It has succeeded in establishing one for electronics in Syria, and another one for energy in Kuwait. The forum for the environment series was originally planned to be established in another of the Maghreb countries (Algeria or Morocco).

In addition to its two main sponsors (MISR and SSRC), the School is currently sponsored by a number of national (such as the Centre National de Coordination et de Plannification de la Recherche Scientific et Technique, Morocco), regional (such as the Arab League Educational, Cultural, and Scientific Organization), and international (such as the United Nations Educational, Scientific and Cultural Organization) scientific, educational, and development organizations.

OBJECTIVES

Through its program, the School aims at reaching the following objectives:

1. Bring Arab scientists up-to-date on the latest scientific developments and state-of-the-art technologies in the field under study.

2. Facilitate direct contact among Arab scientists, and between Arab scientists and their international counterparts to create a favorable atmosphere for scientific and technological cooperation.
3. Make Arab scientists working abroad aware of their home countries' resources, and encourage them to contribute to their scientific and technological development.
4. Provide the international scientific community with an overview of the scientific activities in the Arab countries through the publication of each School session proceedings, which cover the scientific and technological developments in the Arab World.

ARAB SCHOOL AUTHORIZED COMMITTEES

GENERAL SUPERVISORY COMMITTEE

A. R. Kadoura
Assistant Director General, Sciences
UNESCO

A. W. Chahid
General Director
Scientific Studies & Research Center

A. Shihab Al-Din
General Director
Kuwait Institute for Scientific Research

SCHOOL GENERAL SECRETARY

Adnan Rifai
Arab School on Science and Technology

SCIENTIFIC PROGRAM COMMITTEE

Amr Armanazi
Institute of Electronics
Scientific Studies and Research Center, Syria

Bachir Brayez
Electronic Industries Branch, Syria

Haytham El-Yafi
Institute of Chemistry and Biology
Scientific Studies and Research Center, Syria

M. Mohammad Ali
Faculty of Mechanical Engineering
University of Aleppo, Syria

SCIENTIFIC CONSULTANT

Patrick O'Connor
Reliability & Maintainability Manager
British Aerospace
Six Hills Way
Stevenage, Herts
U.K.

LECTURERS

Ronald N. Allan
University of Manchester
U.K.

Robin R. Britton
Plessey Assessment Services Ltd., U.K.

David Hutchins
DHA Ltd.
U.K.

David Newton
Smiths Industries
U.K.

Nihal Sinnadurai
BPA Consultants Ltd.,
U.K.

ORGANIZING COMMITTEE

H. Nahas Armanazi
Session Coordinator
Arab School on Science and Technology

G. Akbik
Administrator, International Relations
Arab School on Science and Technology

A. Kouatly
Administrator, Local Arrangements
Arab School on Science and Technology

Adnan Bach
Public Relations Director
Scientific Studies and Research Center

PREVIOUS SESSIONS	DATE		LOCATION
Solid State Electronics	Summer	1978	Syria
Communications	Summer	1979	Syria
Minicomputers, Microprocessors, and Their Applications	Summer	1980	Syria
Power Systems: Analysis and Planning	Winter	1981	Kuwait
Control Systems: Theory and Applications	Summer	1981	Syria
Solar Electricity Systems	Winter	1982	Kuwait
Pollution and Protection of the Water Quality	Summer	1982	Syria
Modern Signal Processing	Summer	1983	Syria
Applied Arab Linguistics and Signal & Information Processing	Fall	1983	Morocco
Technology Transfer and Adaptation in the Arab World	Fall	1983	Syria
Informatics & Applied Arabic Linguistics	Summer	1985	Syria
Atmospheric Pollution	Summer	1985	Syria
Sensors	Summer	1986	Syria
Reliability Engineering (this book)	Summer	1986	Syria

FORTHCOMING SESSIONS

		LOCATION
Office Automation/Arabization of Databases	1987	Syria/Kuwait/UAE
Marine Pollution	1987	Syria/Kuwait/Tunisia
Production Engineering	1988	Syria
Optoelectronics	1988	Syria
Microwave Applications	1989	Syria
Vaccins, Immunity and Infectuous Diseases	1989	Syria

1
■
Introduction

P. D. T. O'CONNOR
British Aerospace
Hertfordshire, UK

DEFINITION OF RELIABILITY

Reliability is formally defined as the probability that an item will survive without failure for a stated period of time, under stated conditions of use.

The definition of reliability as a probability implies that measurements or forecasts of reliability must be based on probability mathematics, and thus on statistics. Probability and statistics do in fact provide the basis for much reliability theory. However, their application to reliability has been exaggerated in the past. The reasons why these disciplines cannot be applied with high credibility will be discussed below.

VARIATION IN ENGINEERING

All processes and materials used in the manufacture of products are subject to variation. For example, the diameters of machined parts, the purity of silicon crystal, and the forms of solder joints all vary. If these variations are sufficiently large, the items concerned might fail in use. For example, if a machined part is too thin or has cracks, or if a transistor is diffused onto a silicon slice containing impurities, or if a solder joint does not provide good electrical continuity, failure can occur. In all of these cases close control of the variation could prevent failure.

The environment in which products must operate (or be stored, maintained or transported) also varies. Typical environmental variations which can affect reliability are temperature, humidity, shock and vibration. If these variations exceed the ability of the item to withstand them, failure will occur, either immediately or as the result of cumulative damage.

Finally, humans, working as producers, maintainers, handlers or operators of products, introduce further sources of variation which can affect reliability. Examples are incorrect assembly or inspection on a production line, incorrect maintenance such as neglecting lubrication, and electrostatic damage to electronic equipment due to handling while electrostatically charged.

Most variability is described mathematically by the statistical normal distribution, since this is the limiting case when a number of processes combine to generate the overall observed variation. Therefore this

1

distribution is the basis for much of the measurement and control methods in production quality control, and it is assumed for many environmental variations. However, it does not always apply, particularly for shock overload conditions such as electrostatic discharge. In particular, variation due to human behaviour cannot be realistically modelled by any general statistical distribution, and each case needs to be considered separately. For example, the proportion of electronic components incorrectly inserted by a production operator depends upon the component type, location, training provided, lighting, the person's character and mood, etc. Humans as users are particularly prone to introducing stresses and other environmental variation which is very difficult to model or predict.

Variations can cause failure when, for example, the combined effects of a weak component and a high stress level are such that the component cannot withstand the stress. Such interactions of variations are more difficult to evaluate than are the effects of single variations, but they are important sources of failure in products. Sometimes there can be multiple combinations of environmental and strength variation which can lead to failure, often unexpected and not anticipated by the designer or those planning the tests of the product.

In reliability, we are concerned with sources of variation which are time dependent. For example, chemical processes such as corrosion and physical processes such as wear and fatigue can result in an item which is initially very reliable becoming less so with age, due to gradual weakening, until the strength is no longer sufficient to withstand the applied stress.

CONTROL OF VARIATION

To ensure that a product will be reliable, we need to control the sources of variation. In principle, we need to:

1. Design the product so that it will withstand the expected environmental variation.

2. Design the product so that the expected variation due to production processes will not create items that will be too weak to withstand the extremes of environmental variation.

3. Control the production processes so that the design requirements are met.

4. Provide sufficient endurance against progressive weakening, by design to withstand fatigue, corrosion, wear, etc., for the expected life in the expected environments. This includes maintenance.

5. Provide for the expected variations in human behaviour, as producers, maintainers and users.

Obviously it is not always practicable to design and produce products which will always be inherently failure-free in all environments. There will be mistakes, and practical considerations of cost, weight, etc. might limit the extent to which total reliability can be assured. However, systematic attention to these points can lead to extremely reliable design and production, at realistic costs. (Cost aspects will be discussed later). The methods by which these objectives can be realistically attained will be presented during this seminar.

RISK

The level of risk involved in the failure of a product can differ enormously, depending upon the type of failure and its consequences. At one extreme is a major failure of a nuclear power plant leading to major radioactive contamination, at the other is the failure of a fuse on a domestic electrical appliance. Perceptions of risk relate to the consequences in terms of cost of failure, effects such as safety, legal and statutory requirements, market forces such as product reputation and competition, and other factors particular to certain products. The perception of risk can also vary between people and at different times. If a producer is very concerned about the reliability of his product, and he communicates this priority to his design and production people, they will have a stronger perception of the need for quality and reliability than if the management did not emphasise them. Therefore it is a fundamental requirement of any quality and reliability programme that senior management sets the right framework of requirements and motivation.

Recent developments in product liability legislation, increased consumer and industrial user expectations for quality and reliability, and international competition, have all strongly influenced the perceptions of risk inherent in new product development. We can all now demand and obtain high technology products of excellent quality and reliability.

HISTORICAL DEVELOPMENT OF RELIABILITY

The discipline of reliability engineering as a formally recognised response to the growing complexity of new systems, particularly in the military area, was developed in the USA in the 1950´s. The US Department of Defence and the electronics industry set up a joint task force, the Advisory Group on Reliability of Electronic Equipment (AGREE) in 1952. The AGREE report recommended that special disciplines must be imposed during the development phases to ensure that designs would be reliable. These included design analyis techniques and formal reliability demonstration testing. AGREE testing, using statistical methods to demonstrate compliance with reliability requirements, became standardised in US Military Standard 785, issued in 1965 to formalise the complete reliability programme approach.

Similar standards have been introduced in the UK and elsewhere. For example, there is now a NATO reliability standard, and reliability standards have been produced for non-defence work, notably by the British Standards Institution (BS 5760) and by organizations such as NASA and large utilities and companies.

Production quality assurance discplines were originated during World War 2, to control the quality of munitions manufacture. These disciplines were primarily in the area of statistical acceptance sampling. Later developments in statistical process control, to keep production variation within prescribed limits, followed.

Recent developments in reliability have tended to reduce the emphasis on quantitative techniques for predicting and demonstrating reliability, with more emphasis being placed on design analysis and on testing to discover potential weaknesses so that improvements could be made early in the development cycle. This approach has always been the one used by the best commercial industries, and the purchasers and developers of large systems,

particularly the military and defence equipment manufacturers, are now
realising that they are more effective than the methods brought into use in
the 1950's after the AGREE report.

In the production quality field, the major recent developments have been the
spread of the idea of "total quality control", involving all functions in the
drive to higher quality, and the quality circles movement, which started in
Japan and is now spreading in the West.

The use of computers has spread to quality and reliability work, with computer
aided design (CAD) being used for reliability analysis and computers linked to
measurement and test equipment to automate statistical process control.

RELIABILITY AS AN EFFECTIVENESS PARAMETER

Reliability is important only in so far as it affects other dependent factors.
The main dependent factors, common to most products, are:

1. Cost. Every failure results in cost, of repair, replacement,
non-availability to generate income, spares, etc.

2. Availability. Reliability is related to availability (in the steady
state) by the expression

$$\frac{MTBF}{MTBF+MTTR}$$

where MTBF=Mean Time between Failures, and
MTTR=Mean Time to Repair.

This is the availability expression for a simple product, in which any failure
results in the product being unavailable. For more complex systems, involving
redundancy, secondary modes of operation, etc., the relationships become more
complex.

Reliability also can affect safety, goodwill, and other factors which are not
readily quantifiable.

The effects of failures on cost, availability, and other factors can be very
large, particularly when all direct and indirect costs are added. These costs
are often underestimated. Therefore reliability is a very important parameter
of most modern products, ranking equal with performance and cost. Surveys
show that most domestic and industrial consumers are now prepared to pay more
for products with good reputations for reliability.

The relationship between the resources expended on reliability and the
cost savings are often depicted as shown in Figure 1. According to this
picture, there is an optimum level of reliability. However, it is extremely
difficult to forecast by how much the reliability, and therefore the costs,
will be affected by particular activities and expenditures. Also, because of
the high real costs of failures, the optimum reliability for most products is
usually the highest that can be achieved by using the best design and
production techniques. Therefore the picture in Figure 1 can be misleading,
and it should never be used for cutting back on reliability activities unless
clear evidence exists that the optimum reliability is in fact less than the
maximum that can realistically be achieved. Figure 2 represents the situation

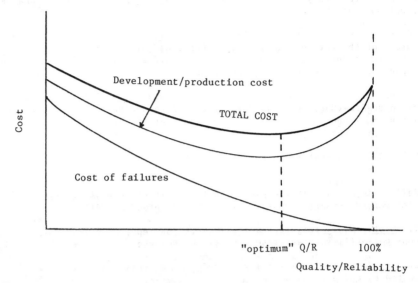

Figure 1   Quality/Reliability and Cost (Old View)

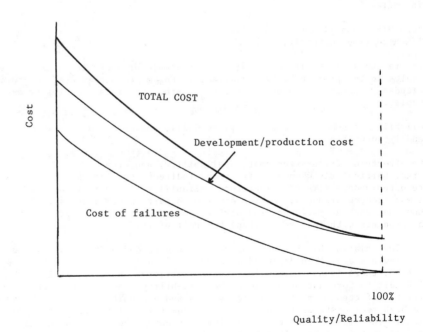

Figure 2   Quality/Reliability and Cost (Modern View)

more realistically. This shows that improving reliability (and production quality) leads to ever decreasing total costs of ownership (and production).

RELIABILITY PROGRAMME ACTIVITIES

The main reliability programme activities are:

1. Determining the requirements and setting up the specifications and motivation.

2. Design analysis methods for reliability.

3. Reliability testing.

4. Production quality control.

These activities should be linked by a data recording and analysis system, incorporating methods for rapid corrective action for design weaknesses and production defects.

Subsequent chapters describe these activities, all of which are directed at reducing variation in the product and at increasing its ability to withstand variation in its operating environment. In turn the reliability and quality activities lead to reduced development, production and in-use costs, improve productivity, and enhance the product's competitive position.

REFERENCES

1. O'Connor, P.D.T., Practical Reliability Engineering, (2nd. edition), J. Wiley, 1985.

2. Deming, W.E., Quality, Productivity and Competitive Position, MIT Press, 1983.

2
■
Reliability Mathematics I

D. W. NEWTON
Department of Engineering Production
University of Birmingham
UK

1. PROBABILITY - SUMMARY

In reliability studies, we are nearly always dealing with
uncertain, or probabilistic, situations rather than deterministic
ones. Reliability itself is usually defined in terms of
"probability of survival".

1.1 Probability Definitions

<u>Classical definition</u>. If an experiment can result in n equally
likely outcomes, ℓ of which can be called "success", then:

Probability of success = $P(s) = \dfrac{\ell}{n}$

This definition has its roots in gambling where it is possible to
assign probabilities 'a priori' (before the event) without
actually performing the experiment.

<u>Frequency definition</u>. This is based on 'a posteriori'
observations of experimental results. If, in an experiment, n
trials result in x successes, then

$P(\text{success}) \rightarrow \dfrac{x}{n}$ as $n \rightarrow \infty$

Reliability usually implies the frequency definition, even if the
value is expressed as a judgment rather than as the result of
actual experiments.

1.2 Rules of Probability

Consider two events, 'A' and 'B':

<u>Complement rule</u>. Denote the event 'not A' = $\overline{A}$
Then $P(A) = 1 - P(\overline{A})$.

<u>Addition rule</u>. $P(A \text{ or } B) = P(A) + P(B) - P(A \text{ and } B)$
If $P(A \text{ and } B) = 0$, A and B are "mutually exclusive", and
$P(A \text{ or } B) = P(A) + P(B)$.

Conditional probability. The probability of event 'A' given that event 'B' has already occurred, is denoted:

$$P(A/B) = \frac{P(A \text{ and } B)}{P(B)}$$

A and B are independent if $P(A/B) = P(A)$ i.e. if the occurrence of event 'B' has no influence on the occurrence of event 'A'.

Product rule. $P(A \text{ and } B) = P(A) \times P(B/A) = P(B) \times P(A/B)$

from which

$$P(B/A) = \frac{P(B) \times P(A/B)}{P(A)}$$

(This is known as Bayes' Rule)

If 'A' and 'B' are independent: $P(A \text{ and } B) = P(A) \times P(B)$

1.3 Discrete and Continuous Event Sets

An event set consists of all the possible outcomes of an experiment. The event set is discrete if the outcomes themselves are discrete - ie they can only take pre-determined values.

e.g. the 52 possible outcomes of selecting a playing card constitute a discrete event set

e.g. the number of failures when 10 items are tested also constitute a discrete event set. In reliability applications, discrete event sets usually consist of numerical values (as in the second example), where they are limited to non-negative integers.

The event set is continuous if the outcomes can take any value in some continuum. In reliability applications we usually refer to a continuum such as operating time which measures the amount of use to which a component or system has been subjected.

2. PROBABILITY DISTRIBUTIONS

2.1 Discrete Distributions

For a discrete event set which has possible outcomes x_1 , x_2 , x_3 etc.

Probability of outcome $x_1 = p_1$
" " " $x_2 = p_2$
" " " $x_i = p_i$, etc

Then $\sum_i p_i = 1.0$

The entirety of the p values constitute the probability distribution, which can be represented graphically as shown in Fig 1.

Mean $= \sum_i x_i \, p_i$ Variance $= \sum_i x_i^2 \, p_i$

8

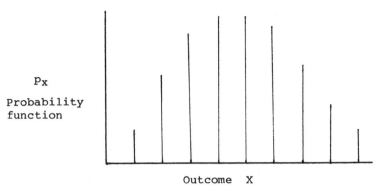

Px

Probability
function

Outcome   X

FIGURE 1 - Discrete Probability Distribution.

2.2   Discrete Distribution Models in Reliability

The most common applications of discrete distributions are to the
situation where the x are observations of the number of failures
in some trial. The most useful models in this application are
the binomial and Poisson distributions.

2.3   Binomial Distribution

This refers to 'Bernoulli Trials'

n items are tested
x failures are observed
(x can take values from 0 to n inclusive)

The probability distribution of x is the binomial distribution:

$$P(x) = \frac{n!}{x!(n-x)!} \, p^x \, (1-p)^{(n-x)}$$

where p is the probability of failure (p = 1 - Reliability)
The mean (expected number of failures) = np.

2.4   Poisson Distribution

When failures occur at random in a continuum of sample space
(usually elapsed time), the process generating the failures is
known as a Poisson process. In the particular case where the
underlying failure rate is constant, this becomes a Homogenous
Poisson Process (HPP) in which the probability of obtaining x
failures in any sample space in which m are expected is given by
the Poisson distribution:

$$P(x) = \frac{e^{-m} m^x}{x!}$$

For a constant failure rate of λ, in a total elapsed time T,
$m = \lambda \, T$.

9

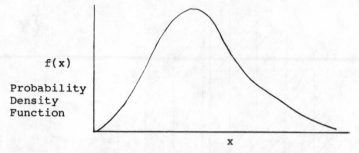

f(x)

Probability
Density
Function

x

FIGURE 2   Continuous Probability Distribution

2.5   Poisson as Approximation to Binomial

For small values of binomial mean np, the Poisson distribution
forms an adequate approximation; i.e. for a Bernoulli trial with
np small (usually <5) and p < 0.1

$$P(x) = \frac{e^{-(np)} (np)^x}{x!}$$

2.6   Continuous Distributions

These describe the entirety of continuous event sets. It is not
possible to specify probabilities at individual sample points.
This can be overcome by dividing the event set into intervals
and assigning probabilities to the intervals - which can be shown
pictorially as a histogram, the total area of which is unity. As
we are describing probabilities rather than data, these class
intervals can be made to tend to zero width - the envelope now
becomes a smooth curve enclosing a total area of unity.

The ordinate is known as the 'probability density function',
f(x).
This function can take many forms within the constraint
f(x)dx =1.0

In reliability applications, the usual application of the concept
is to the modelling of lifetime distributions, as follows.

2.7   Lifetime Distributions

The measure of the reliability of an individual component is its
'lifetime' - the 'time' elapsing between its start of life and
the 'time' at which it fails. The quotes (' ') surrounding 'time'
are intended to convey the idea that whilst it is usual to talk
in terms of that variable, and use the symbol 't', it does not
necessarily imply the passage of 'clock time'. It represents any
suitable measure of a component usage, and it is a matter of
engineering judgment to choose the right one. It might, quite
often, represent a straightforward elapsed time. Alternatively
it would represent:

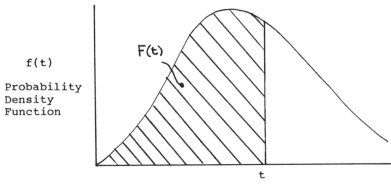

f(t)

Probability
Density
Function

FIGURE 3 - Life-time Distribution.

- operating time (for an application where   time during which
the component does not include any failure inducing aspects),
- operating plus 'stand-by' time
- distance covered (many - but not all - motor vehicle component
failures are km dependent rather than age dependent).
- number of missions (eg in aircraft, the number of flights is
usually the most relevant variable).
- throughput volume (of chemicals, etc, or of gases in petro-
chemical and other process industries).

There are no absolutes in terms of right or wrong choice for a
particular product, as any sensible choices will be closely
correlated. However, the use of the variable that most closely
corresponds to the failure mechanism at work will, in any
particular data set, minimise the uncertainty in parameter
estimation as described later. The value of t at which failure
occurs is, of course, unknown in advance - it is a random
variable that necessitates a probabilistic rather than
deterministic approach. If failures were predictable, as in the
case of the oft-quoted "One Hoss Shay" of Oliver Wendel Holmes
(that lasted for 'a year and a day') then the subject would
become trivial. The key to the modelling of components' lifetime
is the concept of the 'lifetime probability distribution' as
shown in Figure 3.

Without, for the moment, ascribing any particular shape to the
distribution, define

f(t) = probability density function such that the total area of
the figure = 1.

At any value of t that we care to define, the probability that
the component has failed at or before this time is the area under
the curve to the left (shown shaded in Figure 3). This area is
referred to as the 'distribution function' denoted F(t).

(Note that   - at t = 0   F(0) = 0
              - at t = ∞ F(∞) = 1)

and, more generally, $F(t) = \int_{0}^{t} f(u)\,du$

11

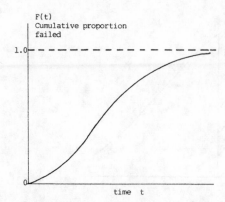

F(t)
Cumulative proportion
failed

1.0

0

time t

FIGURE 4 - Distribution Function.

The graph of F(t) against t depends on the shape of the
probability distribution, but will be of the general form shown
in Fig 4.   We will make particular use of this function when
considering probability plotting and related subjects, as in
"Reliability Mathematics 2".

There are two other related functions that are useful in
describing component reliability:

(a)   Reliability function R(t). This is the probability that a
component has survived to time t, and is simply the complement of
the distribution function, i.e. R(t) = 1 - F(t).

(b)   Hazard function h(t). This is a very useful intuitive
measure of component behaviour, and is defined at rate (ie
probability per unit time). At time t, the hazard function h(t)
is the probability of failure per unit time given survival to
time t.

Let event A be "component fails in interval t to t+δt".
Let event B be "component survives to t".

Then P(A/B) = $\dfrac{P(A \text{ and } B)}{P(B)}$

$= \dfrac{f(t)\delta t}{R(t)}$

ie h(t)   = $\dfrac{f(t)}{R(t)}$

In terms of Figure 1, it is the ordinate at t divided by the area
to the right. Typical behaviour of the function is shown in
Figure 5.

For some components, the hazard function may assume a more-or-
less constant value, ie the likelihood of a failure is
independent of the age of the component. This is often true in

12

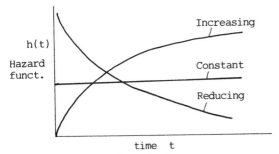

FIGURE 5 - Hazard Function

the case of electronic components and other components where
failures are due to random causes unrelated to component age.
Such failures are themselves 'random' in the strict statistical
sense - but beware of terminology in use that uses 'random' to
denote failures that are rare or unexplained, but not necessarily
of constant hazard. Constant hazard is widely assumed unless
there is evidence to the contrary (it has the attraction of being
mathematically much simpler than the alternatives). The
alternatives are:-

(a) Increasing Hazard Function: the component gets more likely to
fail as it gets older. This will occur in any situation where
use of the product degrades it, for example corrosion, wear,
fatigue etc. As this applies to many engineering components it
suggests that the assumption of constant hazard is, in many
circumstances, at the least questionable.

(b) Reducing Hazard Function: the component gets less likely to
fail as it gets older. A common manifestation of this is the
component that is initially highly stressed due to mis-alignment
and the stress is reduced as the component 'beds in'.

A common misconception is to confuse reducing hazard (which is
fairly rare) with reducing system failure rate (which is
extremely common). The former is a result of individual
components getting better, the latter is due to components of a
particular reliability standard being replaced, on failure with
components that are, on average, better. This may be as a short
term 'burn in' process (where weak components fail early) or on a
longer term 'growth' basis (as described in Reliability
Mathematics 2). The important distinction is that reducing
failure rate can occur with components of reducing, constant, or
increasing hazard function.

A further related function that does not have any obvious
intuitive meaning but which we shall also find useful in plotting
methods for data analysis is that of cumulative hazard, denoted
H(t). This is simply the area under the hazard curve, as shown
in Figure 6.

13

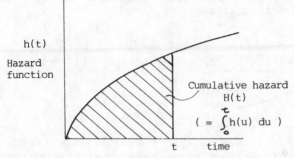

Figure 6   Cumulative Hazard

2.8 Some Lifetime Distribution Models

(for illustrations, refer to fig 2.11 (pp44-5) of Ref 1)

<u>Exponential</u> <u>distribution</u>. This is the lifetime. probability distribution for components subject to constant hazard:

$$f(t) = \lambda e^{-\lambda t} \qquad F(t) = 1 - e^{-\lambda t}$$

$$h(t) = \frac{f(t)}{1-F(t)} = \lambda \text{ (hazard function)}$$

Mean $= \mu = \frac{1}{\lambda}$

The exponential also refers to homogeneous Poisson processes – specifically, in the reliability context, to the stationary point process describing a repairable system. For a system with constant failure rate λ operating over total time T:

Probability of x failures

$$=P(x) = \frac{e^{-\lambda T}(\lambda T)^x}{x!} \text{ (ie Poisson Distribution)}$$

and the distribution of times between successive failures:

$$= f(T) = \lambda e^{-\lambda T} \text{ (ie exponential)}.$$

In this application:

$\lambda = $ failure rate (constant)

$\frac{1}{\lambda} = \Theta = $ "MTBF" (Mean time between failures)

It is important to keep separate these two applications of the exponential distribution – this is covered further in sections 4.1, 4.2 and 4.7 of "Reliability Mathematics 2".

<u>Normal distribution</u>. This is an increasing hazard model with the well-known symmetrical 'bell-shaped' density function. It is of limited use as a lifetime distribution, but is of major importance in the broader field of statistical analysis in general, due to the central limit theorem under which sample

14

averages tend to normality as sample size increases, irrespective of the form of the 'parent' distribution.

Weibull distribution. This is a generalisation of the exponential distribution:

i.e. $F(t) = 1 - e^{-\left(\frac{t}{\theta}\right)}$ (exponential)

$F(t) = 1 - e^{-\left(\frac{t}{\eta}\right)^{\beta}}$ (Weibull) $(0 < t < \infty)$

η is a scale parameter (known as the "characteristic life").
The additional exponent, β, is a shape parameter that changes the shape of the distribution compared with the exponential (for which $\beta = 1$ and $\eta = \theta$)

For $\beta < 1$, the hazard function is reducing (ie the item gets less likely to fail as it gets older)

For $\beta = 1$, the distribution simplifies to the exponential (constant hazard).

For $\beta > 1$, the hazard function is increasing.

More specifically,

for $2 > \beta > 1$, the hazard is increasing at a reducing rate

for $\beta = 2$, the hazard is increasing at a constant rate.

for $\beta > 2$, the hazard is increasing an an increasing rate.

for $\beta = 3.44$, the distribution becomes near symmetrical (and is a close approximation to a normal distribution with mean = 3 x standard deviation).

for > 3.44, the distribution becomes right-skewed.

It is unusual to observe β values greater than around 4.

For large values of β it is often plausible to model a further generalisation of the Weibull for which the probability of failure is zero, in the interval t=0 to t=γ (effectively the origin is moved to t=γ).

     is known as the 'location parameter' and

and $F(t) = 1 - e^{-\left(\frac{t-\gamma}{\eta}\right)^{\beta}}$

The Weibull has many attractions as a lifetime distribution model and has accordingly found widespread application. Among its advantages are:

(i)   Particular suitability for graphical methods (as in

"Reliability Mathematics 2").

(ii) Although originally derived purely empirically, it is theoretically justified under "random shock" theory - it is a log-extreme value distribution.

(iii) Practical experience of many authors has shown that it will nearly always provide a statistically superior fit to valid data sets than any of its competitors.

Whilst caution should be exercised against over-enthusiastic use of the Weibull model (see, for example, Gorski(2)), its popularity is, in general, justified. It will, therefore, be used as the basis for explanation of the reliability data analysis techniques in "Reliability Mathematics 2". For a wider view of probability models and their application in reliability enginering see O'Connor (1) or Mann et al (3).

3. AN INTRODUCTION TO ESTIMATION AND CONFIDENCE INTERVALS

Reliability analysis often consists of estimating the parameters of assumed probability distributions from sample data. These notes will be restricted to point and interval estimates for the mean.

The upper $(1-\alpha_u)$ confidence limit for the mean is defined as the value of the mean μ_u such that the probability of obtaining the observed x, or lower, is α_u.

The lower $(1-\alpha_L)$ confidence limit is similarly denoted μ_L and is such that the probability of obtaining the observed x, or greater, is α_L.

If we state that the true mean is at μ_u or below, there is a probability $(1-\alpha_u)$ that the statement is correct.

If we state that the true mean is at μ_L or above, there is a probability $(1-\alpha_L)$ that the statement is correct.

If we state that the mean is between μ_L and μ_u, there is a probability $(1 - (\alpha_u + \alpha_L))$ that the statement is correct.

If $\alpha_u = \alpha_L = 0.5\alpha$ μ_L to α_u is the $(1-\alpha)$ confidence interval.

3.1 Bernoulli Trials

Test n items.   Observe x failures.

Estimate of probability of failure $= \hat{p} = \dfrac{x}{n}$

Upper $(1 - \alpha_u)$ confidence limit is the solution for   of

$$\alpha_u = \sum_{i=0}^{x} \frac{n!}{i!(n-i)} p_u^{i} (1-p_u)^{(n-i)}$$

16

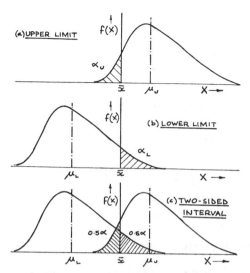

FIGURE 7   Confidence Limits for Mean

Lower $(1 - \alpha)$ confidence limit is the solution for p of

$$\alpha = \sum_{i=0}^{x} \frac{n!}{i!\,(n-i)!}\, p^{i}\, (1-p)^{(n-i)}$$

Solution of these expressions is difficult - rigorous solution requires either an iterative solution or use of tables of the Beta distribution as a conjugate of the binomial (eg Table 16 in Biometrika(4)). For the particular case of $\alpha = 0.05$ and $\alpha = 0.95$, use can be made of the 5% and 95% rank tables in O'Connor (1) (Appendix 6).

For p_u , enter the 95% table with $j = x + 1$.
For p_L , enter the 5% table with $j = x$.

For other values of α, adequate approximations can be obtained from any of the following:
(a)   Interpolation in tables of the cumulative Binomial Distribution.
(b)   Use of a Binomial Nomogram (see O'Connor (1) p.31).
(c)   Use of Poisson approximation as in section 3.3 below.

Example: 20 items are tested. Find the upper 95% confidence limit for p :

(i)   For zero failures observed
(ii) For one failure observed.

(i) In this case, the binomial formula reduces to $p_u = P(0)$

$= (1-p_u)^{n}$

for p_u = 0.05 and n =20,

p_u = 1 $-(0.05)^{0.05}$ = 0.139

This result can be confirmed using any of the other methods

(ii) From O'Connor, appendix 6

j = 2 n = 20 p_u = 21.611% (0.216)

From binomial nomogram:

C = 1 P(c) = 0.05 n = 20 giving p_u = 0.22

3.2 Confidence Intervals for Poisson Process

In observations from an HPP, let

T = Total sample space (time)
x = Observed number of failures

Estimated failure rate (λ) = x/T
Estimated MTTF (θ) = T/x

(a) Upper Confidence Limit for λ (λ_u)

Putting $m_\nu = \lambda_u T$,the value of m_ν (and hence λ_ν) is obtained by the solution of the Poisson probability formula:

$$\alpha_\nu = \sum_{i=0}^{x} \frac{e^{-m_\nu} m_\nu^{i}}{i!}$$

As this equation is tedious to solve (requiring an iterative solution), it is more usual to re-state it in terms of the variate relationship between the Poisson and chi-square distributions. The probability of x or fewer failures in an HPP when we expect m is equal to the probability that a chi-square variate with 2 (x + 1) degrees of freedom exceeds 2m, ie the shaded areas in Figure 8 are both equal to α_ν.

The upper α percentage point of the chi-square distribution (ie the right hand distribution in Figure 8) is extensively tabulated (eg O'Connor (1), Appendix 6) as $\chi^2_{\alpha, \nu}$

where $\alpha = (1-\alpha_u)$ and ν = degrees of freedom.

so, putting $\nu = 2(x+1)$, $\alpha = (1-\alpha_u)$ and $m_u = \lambda_u T$

$$\lambda_\nu = \frac{\chi^2_{(1-\alpha_\nu), 2(x+1)}}{2T} \qquad \text{\{1A\}} \qquad \theta_L = \frac{2T}{\chi^2_{(1-\alpha_\nu), 2(x+1)}} \qquad \text{\{1B\}}$$

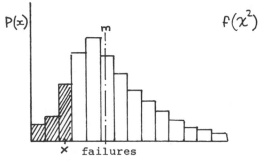 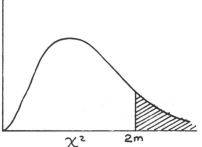

Poisson distribution of mean m   Chi-square distribution with
                                   2(x+1) degrees of freedom.

FIGURE 8   Relationship between Poisson and Chi-square distributions

The foregoing has assumed that the number of failures x is a
random variable occurring in an arbitrary (but fixed) sample
space consisting of the aggregate test time T. In the case where
the test is deliberately terminated on the occurrence of a
failure, the reverse is now true - the random variable is the
elapsed time up to that particular failure. The effect of this
is now to modify equations (1A) and (1B) by reducing the degrees
of freedom by 2.

i.e. for a failure terminated test only:

$$\lambda_U = \frac{\chi^2_{(1-\alpha_U), 2x}}{2T} \qquad \{1c\} \qquad \Theta_L = \frac{2T}{\chi^2_{(1-\alpha_U), 2x}} \qquad \{1D\}$$

(b)   Lower Confidence Limit   for λ (λ_L)

This is rarely specified in practice (it is more usual to need to
know how bad the situation might be rather than how good it might
be), but is included here for completeness.
Referring to Figure 8, we now need to solve:

$$\alpha_L = \sum_{i=x}^{\infty} \frac{e^{-m_L} m_L^i}{i!}$$

ie   we now have $\alpha = \alpha_L$ and replace x by (x+1) giving

$$\lambda_L = \frac{\chi^2_{\alpha_L, 2x}}{2T} \qquad \{2A\} \qquad \Theta_U = \frac{2T}{\chi^2_{\alpha_L, 2x}} \qquad \{2B\}$$

N.B. Care needs to be taken if other tabulations are used - eg
the "α" in Biometrika tables is the complement of that used in
O'Connor.

3.3 Poisson Approximation to Binomial

Using the approximation described in 2.5, combined with the results in 3.2

$$P_u \simeq \frac{\chi^2_{(1-k_u), 2(x+1)}}{2n} \qquad\qquad P_L \simeq \frac{\chi^2_{\alpha_L, 2x}}{2n}$$

For the example in 3.1

For zero failures,

$$P_u \simeq \frac{\chi^2_{0.95, 2}}{40} = \frac{5.99}{40} = 0.133$$

For one failure

$$P_u \simeq \frac{\chi^2_{0.95, 4}}{40} = \frac{9.49}{40} = 0.237$$

(The errors compared with the 'exact' values of 0.139 and 0.216 result from the Poisson being an approximation in this application.)

3.4 Confidence Intervals for Continuous Distributions

The simple discrete distributions considered so far have only one parameter subject to uncertainty – for the binomial, m for the Poisson. The results for the Poisson distribution are directly applicable to the exponential - the single parameter (λ) is common to both. The other continuous distributions described in 2 all have at least two parameters. Problems arise when there is uncertainty in both. These are beyond the scope of this introductory section, but there are also compensating simplifications arising from the central limit theorem. Under this theorem, as sample size increases, the sampling distribution of sample averages will tend to a normal distribution irrespective of the distribution of the variate being sampled.

This gives the simple results:

(i) For known variance

$$\mu = \bar{x} \pm \frac{u_\alpha \sigma}{\sqrt{n}}$$

where σ^2 =   (known) variance
       n =   sample size
       $\bar{x}$ =   observed mean
       u_α =   standardised normal deviate for tail area
($u_\alpha = 1 - \Phi(z)$ as tabulated in O'Connor(1) Appendix 1)

(ii) For unknown variance:

$$\mu = \bar{x} \pm \frac{t_\alpha s}{\sqrt{n}}$$

where t_α is the α percent point of the t distribution with (n- 1) degrees of freedom.
These expressions will apply to the means from complete (ie uncensored) and sufficiently large samples - five or more is usually adequate, but extreme skewness will necessitate larger samples. For the special case of the exponential distribution, the results in 3.2 can be used. For the Weibull distribution, confidence limits for shape parameter are given in Figure 3.7 of Ref (1). For a detailed exposition of point and interval estimation in reliability analysis, see Mann et al (2).

References

1. O'Connor, P.D.T. Practical Reliability Engineering 1986, J.Wiley, London

2. Gorski, A.C. Beware the Weibull Euphoria I.E.E.E. Trans on Reliability, Vol R-17 No 4, Dec 1968, pp 202-3.

3. Mann, N; Schafer, R and Singpurwalla, N.D.. Methods for Statistical Analysis of Reliability and Life data. 1974, J. Wiley, New York.

4. Pearson, E.S. and Hartley, H.O. Biometrika Tables for Statisticians. 1962, Cambridge University Press.

3

■

Reliability Mathematics II

D. W. NEWTON
Department of Engineering Production
University of Birmingham
UK

1. PROBABILITY PLOTTING

1.1 Introduction

This is one of several possible methods   for estimating the
parameters of an assumed lifetime distribution (as discussed in
Paragraph 2.7 of Reliability Mathematics 1).   That this task is
being undertaken automatically makes the assumptions:

(i) That the item involved only fails once (ie it is a component
as distinct from a repairable system).

(ii) That the item lifetimes are all independent observations
from a single distribution with unique parameters ("I.I.D").
(The implications of this assumption will be examined later, in
Section 4.7).

A probability plot is of the sample distribution function, as in
Figure   1.

Modelled distribution functions are fitted to these plotted
points (as shown) - the estimated population C.D.F. is that which
gives the 'best' fit. To facilitate this 'fitting' process:

(i) Assumptions are made about the form of the C.D.F., usually
in the form of restricting   the possibilities to one particular
model.

(ii) Having decided on this model, the plotting axes are
transformed to give a linear plot.

1.2 Basic Procedure

Data is analysed as if it consisted of a life test on a number
(n) of similar components that were tested until k of them had
failed (where k< n).   Even if, as is usual, that is not the way
in which the results were obtained, the procedure operates as
if this were the case and is as follows:

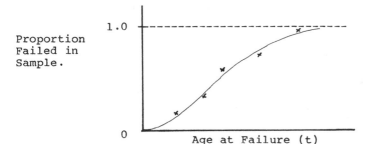

FIGURE 1   Sample Distribution Function

(1) Write down the k failure ages of the components in increasing order of magnitude, viz:

t_1 = smallest ordered age at failure
t_2 = 2nd smallest ordered age at failure

t_i = i'th ordered age at failure

t_k = largest ordered age at failure

(2) At each t_i , calculate the corresponding estimate of the distribution function $F(t_i)$.

Care must be exercised in the calculation of this value, particularly with small values of n. It is incorrect to use the "obvious" values, i/n. This will give an over-estimate, due to the fact that the observation was not at a random value of t, but at a value where it has just increased to a local maximum. Various alternatives have been proposed to i/n, including:

$\dfrac{i}{n+1}$ (mean rank)

$\dfrac{i - 0.5}{n}$ (sample symmetrical C.D.F.)

Median Rank - This is the 50% point of the binomial distribution (see Reliability Mathematics 1, Para 3.1).   Tables are given in Appx 6, O'Connor (1), or it can be approximated by Benard's expression, $(i-0.3)/(n+0.4)$.

The Weibull distribution has particular suitability to probability plotting, and it will be used for illustration in the rest of this section.

This procedure is applicable to data that is either complete (i.e. all the items in the sample have failed, and n = k) or 'singly censored' data where all (n - k) unfailed items have survived to times equal to or greater than t_k. If these

23

conditions are not met, it will be necessary to use a modified analysis as described in Section 3.

2. WEIBULL ANALYSIS

2.1 The Weibull C.D.F. is given by

$$F(t) = 1 - \exp\left(-\frac{t}{\eta}\right)^{\beta}$$

re-arranging:

$$\frac{1}{1 - F(t)} = \exp\left(\frac{t}{\eta}\right)^{\beta}$$

$$\ln\frac{1}{1 - F(t)} = \left(\frac{t}{\eta}\right)^{\beta}$$

$$\ln\ln\frac{1}{1 - F(t)} = \beta\ln t - \beta\ln\eta$$

This represents a straight line of form $y = mx + c$

Where vertical ordinate $y = \ln\ln\left(\dfrac{1}{1 - F(t)}\right)$

horizontal axis $x = \ln t$

slope $m = \beta$

intercept on y axis $c = \beta\ln\eta$

Thus, if we plot the natural logarithm of experimental values of time to failure on the horizontal axis and

$$y = \ln\ln\left(\frac{1}{1 - F(t)}\right)$$

on the vertical axis, if the observations come from a Weibull distribution, we will obtain a straight line. Furthermore, we can estimate the shape parameter from the slope of the line, and the scale parameter (characteristic life) from the fact that the vertical intercept = $-\beta\ln\eta$, or, more simply, the horizontal intercept = $\ln\eta$.

This will permit plotting of the Weibull distribution function on linear graph paper - an alternative (and more popular) method is to transform the plotting axes (rather than the data) as follows:

$$y = \ln\ln\left(\frac{1}{1 - F(t)}\right)$$

$$e^{y} = \ln\left(\frac{1}{1 - F(t)}\right)$$

$$-e^{y} = \ln(1 - F(t))$$

$$F(t) = 1-e^{(-e^y)} \quad (=G(y))$$

and $\quad x = \ln t$, giving $\quad t = e^x$

Various commercial plotting papers are available that incorporate this transformation. The version used in the following example is that due to Nelson(2), the plot being shown in figure 2.

2.2   Example

The data in Table 1 below refers to a life test on a sample of 20 switches. The quoted life is the number of operations at which failure occurred. The unfailed switches were removed after 4500 operations.

(Note that the sample size n is 20 and not 15 (the number of failures)). This data is 'single censored' in that the test was stopped at an arbitrary time before all the switches had failed. There is no particular difficulty in handling this data because all the censoring times (ie times at which unfailed items are removed) are greater than the largest time to failure. When this is not the case we have 'multiply censored' data that is much more difficult to deal with.)

TABLE 1. Example - Cycles to failure of switches

Failure Number (i)	Number of Operations (t)	F(t) (Median Rank)
1	430	0.034
2	900	0.083
3	1090	0.131
4	1220	0.181
5	1500	0.230
6	1910	0.279
7	1915	0.328
8	2250	0.377
9	2600	0.426
10	2610	0.475
11	3000	0.525
12	3390	0.574
13	3430	0.623
14	3700	0.672
15	4050	0.721

Referring to figure 2,

Shape parameter. Construct a perpendicular from the fitted line to pass through the 'estimation points' in the top left hand corner of the paper. Read off the estimate ($\hat{\beta}$) where this perpendicular crosses the scale.
In this case $\hat{\beta} = 1.62$.

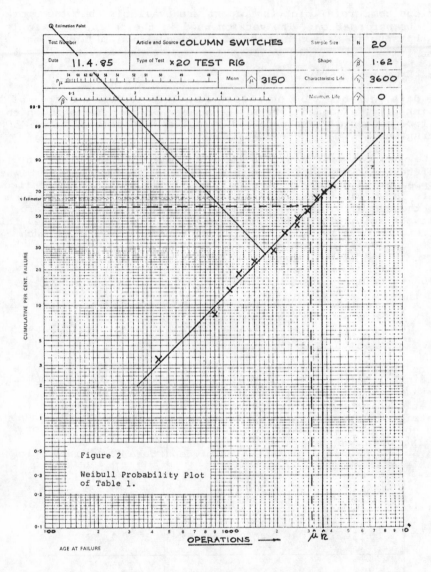

FIGURE 2. Weibull plot of switches data.

26

Characteristic life. Find the point on the fitted line where the 'cumulative percent failure' value is 63.2. (This is printed on the paper as a broken line marked ' η estimator'). From this point, drop a vertical line to the t axis. This t value is the characteristic life.
In the example, $\hat{\eta}$ = 3600 operations.

Mean life. Find the estimate of the percentage failed at the mean life (P(μ)) as the point where the "perpendicular" line crosses the P(μ)scale.
In the example, $P(\mu)$ = 57%.

Identify this value of P(μ) on the vertical axis, read across horizontally to the plotted line and then drop a vertical to the horizontal 't' axis. This 't' value is the mean life.
In the example, $\hat{\mu}$ = 3200 operations.

Standard deviation. This is given (approximately) by multiplying the characteristic life by a factor 'B' as tabulated below (interpolating where necessary).

β: 1.0 1.5 2.0 2.5 3.0 3.5 4.0 5.0
B: 1.0 0.61 0.51 0.37 0.32 0.28 0.25 0.21

NOTE.   The mean is given by: $\mu = \eta\left(\Gamma\left(1+\frac{1}{\beta}\right)\right)$

and the standard deviation by: $\sigma = \eta\left(\left(\Gamma\left(1+\frac{2}{\beta}\right)\right)-\left(\Gamma\left(1+\frac{1}{\beta}\right)\right)^2\right)^{\frac{1}{2}}$

2.3   Some Common Problems in Weibull Analysis

There is a temptation to be over-critical of departures from a straight line in a Weibull plot. It must, however, be remembered that we are not dealing with a deterministic physical relationship (such as a "Hookes Law" plot) but a statistical distribution function in which errors are unavoidable. Indeed, too good a fit indicates that the data could be suspected of having been 'adjusted'.
There are, however, two examples of departure from linear plots that sometimes occur in practice.

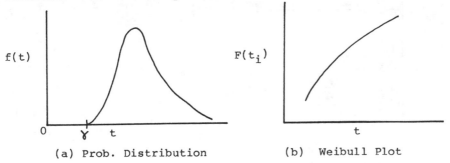

f(t)   F(t_i)

0   γ   t   t

(a) Prob. Distribution   (b) Weibull Plot

FIGURE 3.   Positive Location Parameter

27

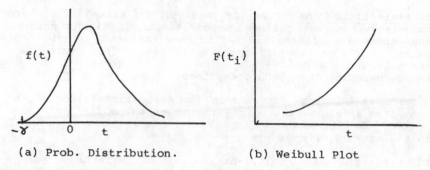

(a) Prob. Distribution.   (b) Weibull Plot

FIGURE 4   Negative Location Parameter

(i) Location Parameter (Monotonic Curved Plot)

The appearance of a monotonic curve indicates that the origin of
the t values is not zero, but some non-zero constant known as
the 'location parameter'. Conventional symbols are γ or t_0 .
A positive value of γ describes a life-time distribution as
shown in Figure 3(a), and will result in a curved Weibull plot as
in Figure 3(b). This situation arises when there is some interval
(from t =0 to t = γ) during which failures cannot occur. An
example might be the erosion of a protective coating. A negative
value of γ describes a lifetime distribution as shown in Figure
4(a) and will result in a curved Weibull plot as Figure 4(b).
This could be caused, for example, by a 'shelf life' problem
where failures can occur before the start of normal service life.

In either case, the existence of a location parameter should not
be accepted without some sensible engineering explanation. It
should also be viewed with suspicion if it is linked with a low
value (less than about 1.5), which would imply a sudden
discontinuity in the distribution.   If there is a plausible
justification, the procedure for estimating γ is as follows:

(1) Draw three equi-spaced parallel horizontal lines on the
plot, as shown in Fig 5. (The 'equispacing' is in terms of

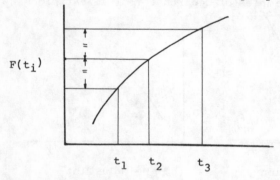

FIGURE 5   Estimation of Location Parameter

28

actual distance on the paper, not the F(t) scale). The distance between the top and bottom lines is not critical, but they should just encompass all the plotted points.

(2) Drop verticals from the inter-section of these lines with the plotted curve at t_1, t_2 and t_3 as shown.

(3) $\hat{\gamma}$ is estimated by:

$$\hat{\gamma} = t_2 + \frac{(t_3 - t_2)(t_2 - t_1)}{(t_3 - t_2) - (t_2 - t_1)}$$

When this estimate has been obtained, it should be subtracted from all the observed failure and censoring times, and the data re-plotted. The re-plot will give a straight line from which the other parameters can be estimated in the usual way.

(ii) Competing Failure Modes ('Bent' Plot)

A plot such as Figure 6 indicates that two independent, competing, causes of failure are present, with different Weibull parameters. What must NOT be done in this case is simply to estimate the β and η values for each part of the line, as if the other did not exist - this will give completely incorrect values. The correct procedure is to produce separate plots for each mode of failure, treating the other one(s) as censorings.

It should be noted that ,even if the dichotomy between the lines is clear, it is difficult to separate the data points referring to each line from the plot alone - and dangerous to try. Identification of the failure modes can only be satisfactorily achieved by engineering investigation of the failed components. Further, the plot cannot be relied upon to demonstrate two modes, even if they are present. Mixing of distributions will always increase the variance of the data, which will reduce the apparent value. Important 'wear-out' modes are often not detected because the data contained other failure modes which obscured them.

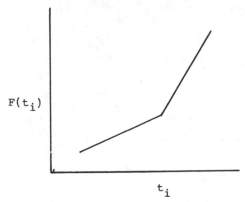

FIGURE 6   Two Competing Failure Modes

3. HAZARD PLOTTING

Section 2 has dealt with the analysis of either complete date (where all the items in the sample have failed) or 'singly censored' data (where all the survival ages are equal to or greater than the largest time to failure). Such data usually arises from life tests on specially procured samples. When we are dealing with the analysis of field service failure data, the problem of multiple random censoring often appears. In this case, unfailed life-times (censorings) are mixed with the failure times, which makes estimation of $F(t)$ less obvious than previously. Consider, for illustration, a sample of five motor vehicles that were tested for a particular mode of failure:

No.1: Currently at 22,000 km without failure
No.2: Failed at 40,000 km. Removed from test
No.3: Failed at 5,100 km. Removed from test
No.4: Destroyed in accident at 9,500 km.
No.5: Failed at 15,000 km. Removed from test.

Putting these events in increasing order we have:

5100	Failure	(F1)
9500	Censoring	(C1)
15000	Failure	(F2)
22000	Censoring	(C2)
40000	Failure	(F3)

Sample size (n) = 5. Failures (k) = 3.

For the first item there is no problem - it is the first failure in a sample of 5.

For the failure at 15000 km there are two possibilities:
(i) It is the 2nd failure (ie C1 fails F2).
(ii) It is the 3rd failure (ie CI fails before F2)

By considering all the arrangements of F2, it is possible to arrive at a 'mean order' of 2.25. Similarly, F3 could be the third, fourth or fifth failure depending on what would have eventually happened to C1 and C2 - its mean order is in fact 4.125.

Calculation of mean order can be tedious with large samples. Other methods have been suggested (eg Kaplan-Meier estimates), but these notes will use the 'Hazard Analysis' due to Nelson (3), a method that has achieved widespread use in Reliability analysis and is particularly applicable to the Weibull distribution. At any time at which a failure occurs, the sample estimate of the hazard function is simply the number of failures at that time divided by the number in the sample available to fail (survivors) immediately before that time. The cumulative sum of these values will give a sample estimate of the cumulative hazard function $H(t)$ as described in Reliability Mathematics 1.

Using the previous example, we have:

Time	No of Fails	Survivors	Hazard	Cumulative Hazard
T	(x)	(s)	h(t) (=x/s)	H(t) (= $\sum$ h(t))
5100	1	5	0.200	0.200
9500	–	4	–	–
15000	1	3	0.333	0.533
22000	–	2	–	–
40000	1	1	1.000	1.533

For Weibull analysis there are now two possibilities:

(i) Plot log H(t) against t - this will, under the Weibull model, give a straight line of slope β .
or
(ii) Convert the H(t) estimates into F(t) using

$$F(t) = 1- e^{H(t)}$$

t	H(t)	F(t)
5100	0.200	0.181
15000	0.533	0.413
40000	1.533	0.784

These F(t) values can now be plotted on Weibull plotting paper in the usual way.

EXAMPLE (All ages refer to SYSTEM hours).

Three repairable systems were monitored for failures of a particular component (of which there are two per system, denoted A and B). The results were as shown below:

System 1
Compt A - failed (and replaced) at 3780 and 6362 hrs
Compt B - failed (and replaced) at 4885 hrs
System currently at 8000 hrs.

System 2
Compt A - failed (and replaced) at 1040 hrs
Compt B - no failures

Both components replaced as precautionary
measures at 5000 hrs
System currently at 6500 hrs

System 3
Compt A - failed (and replaced) at 2180 hrs
Compt B - failed (and replaced) at 2777 and 5082 hrs
System currently at 6950 hrs.

The data is re-organised in terms of component lives and presented in Table 2. Note that component censorings occur at the current system lives (as they are still operating and unfailed) and also at the planned replacements in System 2.

In Table 2:

Column 1 shows the event identification (eg 3B2 refers to the second event for component B in system 3.

Column 2 shows the corresponding component age.

Column 3 identifies whether the event is a failure or a censoring.

Column 4 is the number of survivors.

Column 5 is the sample hazard.

Column 6 is the cumulative sum of Column 5. (i.e. the cumulative hazard $\overline{H}(t)$.

Column 7 is the sample distribution function (from $F(t) = 1 - e^{-H(t)}$)

TABLE 2. Data for cumulative hazard example

1	2	3	4	5	6	7
2A1	1040	F	15	0.0667	0.0667	0.0645
2A3	1500	C	14	-	-	-
2A2	1500	C	13	-	-	-
1A3	1638	C	12	-	-	-
3B3	1868	C	11	-	-	-
3A1	2180	F	10	0.1000	0.1667	0.1535
3B2	2305	F	9	0.1111	0.2778	0.2426
1A2	2582	F	8	0.1250	0.4028	0.3316
3B1	2777	F	7	0.1428	0.5456	0.4205
1B2	3115	C	6	-	-	-
1A1	3780	F	5	0.2000	0.7456	0.5256
2A2	3960	C	4	-	-	-
3A2	4770	C	3	-	-	-
1B1	4885	F	2	0.5000	1.2456	0.7122
2B1	5000	C	1	-	-	-

These results can be plotted on Weibull probability paper from which the Weibull parameters are estimated as:

Shape Parameter $\hat{\beta}$ = 2.0
Scale Parameter $\hat{\eta}$ = 4200
(Mean $\hat{\mu}$ = 3750)

The second example in Table 3 shows lifetimes of a component in the engine of a Diesel truck. It includes failures and planned replacements of the component and current unfailed ages (on trucks that are currently operating). These have been re-arranged as in the previous example to give ordered events (in terms of distance covered (km) at failure). The only difference from the previous example is that the data is grouped into class intervals. The simplifying assumption is made that failures and censorings are spread evenly throughout each interval - from this assumption we plot the km values as the mid-point of each interval, and determine the hazard from the average of the survivors at the beginning and the end of each interval (eg the hazard in the class interval 25-30 is 3/46 = 0.0652) the survivors at the beginning and the end of each interval (eg the hazard in the class interval 25-30 is 3/46 = 0.0652).

In Table 3:

Column 1 defines the class interval (in thousands of kilometres)

Column 2 is the number of failures in the interval.

Column 3 is the number of censorings in the interval.

Column 4 is the number of survivors.

Column 5 is the sample hazard.

Column 6 is the cumulative sum of Column 5. (i.e. the cumulative hazard $\overline{H}(t)$).

Column 7 is the sample distribution function (from $F(t) = 1 - e^{-H(t)}$).

TABLE 3. Water pump failure data.

1	2	3	4	5	6	7
0- 5	0	1	68	0	0	0
5-10	2	4	67	0.0313	0.0313	0.031
10-15	3	3	61	0.0517	0.9830	0.080
15-20	2	2	55	0.0381	0.1211	0.114
20-25	1	1	50	0.0204	0.1415	0.132
25-30	3	1	48	0.0652	0.2067	0.187
30-35	3	0	44	0.0706	0.2773	0.242
35-40	1	3	41	0.0256	0.3029	0.261
40-45	1	7	37	0.0303	0.3332	0.283
45-50	0	2	29	–	–	–
50-55	1	4	27	0.0408	0.3740	0.312
55-60	1	7	22	0.0555	0.4296	0.349
60-65	0	6	14	0	–	–
65-70	0	3	8	0	–	–
70-75	2	1	5	0.5714	1.0010	0.632
75-80	0	1	2	0	–	–
80-85	0	1	1	0	–	–
>85	0	0				
	--	--				
	20	48				

The probability plot in this instance will be of F(t) against the mid-point of each interval (ie 2500, 7500 etc.).

4. POINT PROCESSES - THE RELIABILITY OF REPAIRABLE SYSTEMS

4.1 Introduction

Sections 1 to 3 have dealt with the reliability of <u>components</u> which are assumed to fail only once. The situation has been modelled with the concept of the 'life-time distribution', with the underlying implication that observed life-times are independent observations from the same life-time distribution with the same parameter values (ie they are independent, identically distributed (I.I.D.) values).

This section deals with repairable systems where the occurrence of a failure does not necessarily cause the 'end of life' of the system; rather, the failed element is repaired or replaced to restore the system to its operating state. On this basis, for such a system we can observe the system elapsed times at which failures occur.

Defining:
y_i = system age at the i'th failure
t_i = interval between (i-1)th and i'th failure

the history of a typical system can be represented as shown in figure 7.

We will make some initial simplifying assumptions:

1. The repairs that take place at each y_i are 'perfect', ie they restore the failed equipment to its "as new" condition.

2. The failures occur at random, but at a constant underlying rate. This implies that the failures are the result of a homogeneous Poisson process (HPP) giving inter-failure times t that are exponentially distributed (and the number of failures in specified time intervals have a Poisson distribution).

(Circumstances when these assumptions are justified, and the procedures to adopt when they are not, are discussed in 4.6 below.)

For a system that complies with these assumptions, a plot of cumulative failures N(y) against system age y would produce a result as shown in Fig 8, the expected relationship being a straight line. The slope of this line $\frac{d\ N(y)}{dy}$ is called the "system failure rate" (usual symbol λ), and its reciprocal is called the "mean time between failures" (MTBF), (usual symbol θ).

34

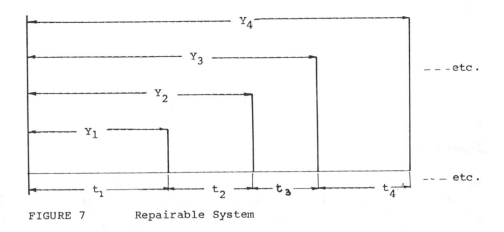

FIGURE 7   Repairable System

4.2 A Note on "Failure Rate".

"Failure Rate" as described above relates specifically to the failures per unit time of a repairable system - it must be distinguished from the concept of hazard functions for non-repairable components that form elements within such a system. Such components may exhibit:

1. Reducing hazard - in which case they become less likely to fail as they get older.

2. Constant hazard - in which case their probability of failure is unaffected by their age.

3. Increasing hazard - in which case they become more likely to fail as they get older.

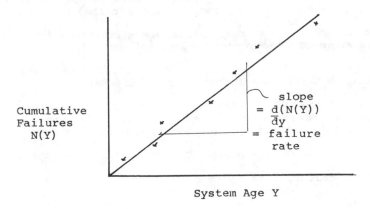

FIGURE 8.   System Failure Rate

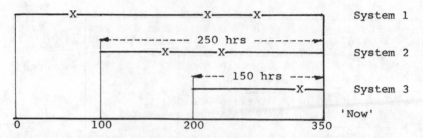

FIGURE 9.   Repairable System Example

The essential point in considering a repairable system is that, providing assumption (i) in Section 2.1 applies, for a complex system the failure rate will tend to a constant value as y increases, irrespective of the hazard pattern. As an example, consider a system consisting of a large number of filament light bulbs. Such bulbs exhibit increasing hazard yet, after an initial settling down period, the rate at which bulbs fail and are replaced will assume an approximately constant value. Confusion has arisen because the term "failure rate" is often used to describe "hazard function" (sometimes modified to "instantaneous failure rate"). Similarly, in another attempt to distinguish between the concepts, system failure rate is referred to by such convoluted terminology as "pseudo failure rate" or "Rate of occurrence of Failures" (ROCOF). In the absence of any agreed terminology, the confusion is something that we have to live with - the important thing is to realise that the concepts are separate. It is perfectly possible to have, for example:

a) Increasing hazard and constant failure rate (as in the light bulbs example)

b)   Increasing hazard and reducing failure rate (in the case of a 'reliability growth' programme for the light bulbs).

c)   Constant hazard and reducing failure rate (as in reliability growth for a system of electronic components) etc.

4.3 Reliability Analysis under Constant Failure Rate

A useful effect of the HPP assumption is that it is possible simply to aggregate system life y over any number of systems, irrespective of system age to give a total elapsed time T.

If, in this elapsed time T, we have observed x failures, then:

$\hat{\lambda}$ = estimated failure rate = x/T

$\hat{\theta}$ = estimated MTBF   = T/x

For example, Fig 9 above represents the history, so far, of three

systems. System 1 starts its life at time 0. System 2 starts
its life when System 1 has been operating for 100 hours. System
3 starts its life when System 1 has been operating for 200 hours
(and System 2 for 100 hours). Failures are denoted -x-.
Currently Systems 1, 2 and 3 have been operating for 350, 250 and
150 hours respectively.

The current TOTAL elapsed time is 350 + 250 + 150 = 750 hours,
during which there have been 6 failures.

The estimated failure rate is therefore 6/750 = 0.008
failures/hour, and the estimated MTBF = 125 hours.

Under the HPP model, over any 'packet' of elapsed time totalling
T, the expected number of failures is   T. The probability of
observing some arbitrary number of failures, x, is given by the
Poisson distribution.

$$P(x) = \frac{e^{-m} m^x}{x!} \qquad \text{(where } m = \lambda T\text{)}$$

Confidence intervals for θ (or λ) can be calculated as explained
previously (Lecture RM1, para 3).

4.4 Reliability Demonstration for a Repairable System

Reliability (usually in the guise of MTBF) is often quoted as a
specification requirement or a system, together with a confidence
level with which it is to be demonstrated that the requirement is
met.

- for example "It shall be demonstrated with a 90% confidence
that the MTBF is not less than 150 hours".

This is achieved by producing a demonstration test result whose
probability of occurrence, if the MTBF is equal to or lower than
the required value, is (1 - confidence level).

- in this example, we need a test result such that, if the MTBF
is 150 hours, the probability of obtaining the number of failures
that in fact occurred, or fewer, is not more than 10%. The test
is defined by T, the accumulated test time, obtained from a re-
arrangement of equation 1B, ie

$$T = K\theta_L \qquad \text{where } K = \frac{\chi^2_{(1-\alpha), 2(x+1)}}{2}$$

K is a factor by which the required MTBF is multiplied to give
the required test time, knowing the required confidence level ($1-\alpha$)
) and the number of failures that have been observed - for
convenient reference, it is shown in table 4 for some
conventional values of confidence level (expressed as
percentages).

37

TABLE 4. Multiplication factors for required M.T.B.F. for
calculating test times for demonstration of achievement at given
confidence level

Confidence Level	Number of failures					
	0	1	2	3	4	5
50%	0.70	1.68	2.67	3.67	4.67	5.67
75%	1.39	2.69	3.92	5.06	6.27	7.42
80%	1.61	2.99	4.28	5.52	6 72	7.91
90%	2.30	3.89	5.32	6.68	7.99	9.27
95%	3.00	4.74	6.30	7.75	9.15	10.60
99%	4.60	6.64	8.41	10.04	11.60	13.11

So, in our example, we need to accumulate 150 x 2.30 = 345 hours
without failure. Should we get a failure before this time, the
test time required inreases to 150 x 3.89 = 5.83 hours. Should a
second failure occur, it further inreases to 150 x 5.32 = 798
hours, and so on.

4.5   Acceptance Tests

The problem with demonstration as described above is that whilst
a reliable system will be accepted, an unreliable one will not be
rejected - the accumulated hours will simply fall progressively
further behind the target as testing continues. For instance, in
our example, we might get one failure in 210 hrs (so we need to
carry on to 583 hours) but we get a second failure at 330 hours
(so we need to carry on to 789 hours) but we get a third failure
at 425 hours (so we need to carry on to 1002 hours) etc. Clearly
this product is so unreliable that it will never meet the
reliability requirements in its present state of development.
Human nature being what it is, there is a temptation to keep on
testing in the hope that the test requirement will be met - to
avoid this, it is useful to introduce a procedure whereby a
'reject' decision can be reached (meaning 'take the system off
test, go away and improve it').

The way to define a suitable acceptance test is to specify two
levels of failure rate, as follows:-

a)   a low value of the failure rate, λ_1 at which we have a
probability of $(1-\alpha)$ of accepting the product.

b)   a higher value of the failure rate, λ_2 at which we have a
probability of β of accepting the product.

   α and β are small values (typically 0.05, but we can make
them any value less than 0.5) known, respectively, as the
'producers' and 'consumers' risks. The producers' risk (α) is

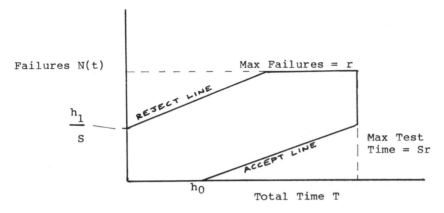

FIGURE 10   Sequential Probability Ratio Test

the risk of wrongly rejecting a product which has a
'satisfactory' failure rate λ_1 .   The consumers' risk (β) is
the risk of wrongly accepting a product which has the
'unsatisfactory' failure rate λ_2. λ_1 / λ_2 is known as the
'discrimination ratio'.

Various testing procedures have been proposed, but a very
efficient and widely used one is the "sequential probability
ratio test (SPRT)".   In this type of test, a log is kept of the
accumulated test time (T) and the total number of failures
(N(T)), and they are plotted as in Fig 2 (except that the test is
not necessarily restricted to a single system, so we use T
instead of y).   The accept/reject criteria are given by drawing
on the plot a pair of parallel lines as shown in Fig 10, of
equations:

$$\text{Accept line - T} = h_0 + S\,(N(T))$$
$$\text{Reject line - T} = -h_1 + S\,(N(T))$$

The design and application of this type of test is covered under
"Reliability Testing 2".

4.6   The Assumption of Constant Failure Rate

The foregoing material has assumed throughout that the failure
rate exhibited by the system (λ) is constant, ie the plot of
cumulative failures against total elapsed time is a straight line
as in Figure 8.

This assumption will, in general, be valid if all the elements in
the system exhibit constant hazard function.   This is generally
true in the case of electronic components but, in the case of
mechanical components, this is not the case - non-constant hazard
functions are commonplace, usually increasing but also including
the "initial rise followed by a fall" typically found in fatigue
failures.   If there is a large number of components, with no
dominant failure mode, there will be a 'mixing' effect of the

39

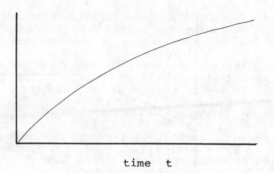

N(t)
Cumulative Failures

time   t

FIGURE 11. Reliability Growth

failures and replacements so that the overall failure rate (which can be approximated by the sum of the individual component hazards) will soon settle down to an essentially constant value.

This constant failure rate pattern now depends on the further assumption of "good as new" replacement of failed components – that is the replacement item will have the same life-time distribution as the item it replaced (which is not the same as saying it will fail at the same age). There are some circumstances in which we expect this not to be true, viz:

i) The consequence of engineering development make the replacement item superior to the failed item. In this case we would expect it to last longer. If this happens to several of the failure modes, we would expect the times between successive failures to tend to increase, giving a result as in Figure 11. This particular form of non-constant failure rate has been the subject of extensive further analysis under the guise of "Reliability Growth Modelling" - see Section 6.6.

ii) Conversely, the replacement item may be 'worse than new'. This usually results from the replacement item having received a less than perfect repair. As an example, a fatigue crack in a casting may be repaired by welding, but other potential cracks are unaffected and their propagation is related to the start of life of the casting, not the repair time. Reliability deterioration of this sort will give a result as in Fig 12.

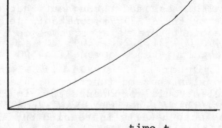

N(t)
Cumulative
 failures

time t

FIGURE 12 - Reliability Deterioration

Reliability growth or deterioration must not be confused with reducing or increasing hazard - they are independent of each other, and any combination is possible as discussed in para 4.2.

A useful statistical test for the hypothesis of constant failure rate (strictly for the hypothesis of a Homogeneous Poisson Process) is due to Laplace - the quantity

$$\sqrt{12n}\left(\frac{\sum_{}^{n} T_i}{n t_o} - 0.5\right)$$

is tested as a standardised normal deviate, where
T_i = total system time at i'th failure
t_o = final total system time
n = number of failures at t_o

(e.g., the hypothesis of HPP is rejected at the 90% significance level if the modulus of this test statistic exceed 1.28. For 95% significance, this value is 1.65 - for other significance levels, values can be obtained from a table of the normal distribution.)

If the test gives a signicant positive value, reliability deterioration is occurring. For a negative value there is a reliability growth.

For a failure terminated test, where the test stops at the moment of the n'th failure (t_n), the expression above is modified to:

$$\sqrt{12(n-1)}\left(\frac{\sum_{i=1}^{n-1} T_i}{(n-1) t_n} - 0.5\right)$$

EXAMPLE
The table below shows the results of a prolonged trial during the development testing of four prototype motor vehicles:

Vehicle Number	Cumulative vehicle kM at which failures occurred (thousands)
1	2,17, 19, 16, 38, 57, 101, 141 (150)
2	0.3, 10, 12, 15, 41, 87, (120)
3	3,5,11,14, 23, 51, 66, 79, 113,(120)
4	0.9, 5, 6, 9, 17, 19, 53, 71, (100)

(figures in parentheses are km at the end of trial for each vehicle).

There are 31 failures at total elapsed times of:
1.2, 3.6, 8, 12, 20, 20, 24, 36, 40, 44, 48, 56, 60, 68, 68, 76, 76, 92, 104, 152, 164, 204, 212, 228, 264, 284, 316, 348, 403, 439 and 481 and the total distance covered is 490 (all in units of 1000 km).

(This assumes all vehicles have a common time origin - effectively that they were in the same state of development at time 0. On this basis, the first failure (in vehicle 2) occurred at t = 0.3, so the total distance was 0.3 x 4 = 1.2. Other values were calculated similarly, making due allowance for the progressive 'dropping-out' of the trial of vehicles 4, 3 and 2.)

41

$$\frac{\sum T_i}{nt_o} = \frac{(1.2 + 3.6 + 8 \ldots\ldots +439 + 481)}{31 \times 490}$$

$$= \frac{4351.8}{31 \times 490} = \underline{0.286}$$

For zero trend, this value would equal 0.5. To convert to the test statistic:

$$\sqrt{12n}\left(\frac{\sum T_i}{nto} - 0.5\right) = 12 \times 31 \ (0.286-0.5) = -4.13$$

so the failure rate is clearly decreasing - reliability growth is present (as one would hope during development testing).

4.7 Reliability Growth Analysis.

For a repairable system, the widespread assumption of constant failure rate (as in 4.3) depends on the assumption of 'good as new' replacements, ie the replacement for a failed element is of the same standard (with the same expected life-time) as the failed item. The actual eventual achieved life-time is, of course, subject to sampling variation, and may be greater or smaller than that of its predecessor - its expected (or mean) value is the same. If this were to be true, it would, in fact, be a criticism of the engineering development process. Every failure has an engineering cause, and in an ideal world the product design should be improved on the basis of information obtained from failures until their causes are eliminated. This ideal world is rarely, if ever, achieved in practice because the law of diminishing returns would make it uneconomic to do so, but what one can hope for is to eliminate many of the more persistent causes of premature failure leaving a 'base' level of unavoidable wear-out and fatigue failures, plus, perhaps, a new 'nuisance' failures of low level and minor consequence.

As such engineering development takes place, failed items will tend to be replaced not by 'good as new' items, but by 'better than new', and the failure rate curve will show a reducing slope, as in Fig. 11. The Laplace test can be used to see if such growth is taking place, with a general rule that the larger the (negative) test statistic, the more rapid the growth.

Some more specific models have been proposed which assume specific forms for the shape of the curve in Fig 11 - the simpler models in most widespread use being based on an exponential 'learning curve'. Reliability growth models are described in detail in "Reliability Testing 2".

4.7 The 'I.I.D.' Assumption.

The discussion of component failure analysis in sections 1 - 3 has made the assumption that the observed t_i are all independent observations from the same life-time probability distribution -

they are "independent and identically distributed" (I.I.D.)
observations. In the simple case of a sample of nominally
identical components subjected simultaneously to a rig test,
there would not be any reason to question this assumption. In
the case of results that come from field service, however, there
are dangers in not doing so. In component analysis, the observed
times are ordered by magnitude without any reference to the
sequence in real time in which they occur. There are
circumstances in which the sequence is important, in which case
the failures are not I.I.D. and simple distribution fitting is
not justified.

Consider, for illustration, the following eight observed failure
times and the circumstances in which they might have arisen:
67, 127, 32, 125, 96, 19, 196, 52. Figure 13 shows four of the
many possible structures which might have been the origins of
this data.

Figure 13(a) represents a 'rig test' with the results being
obtained from simultaneous systems. Figures 13(b), (c) and (d)
represent some possible outcomes from the instalation of the
component in a repairable system. Figures 13(c) and 13(d) show
extreme examples of the sequencing of the failures having a
considerable affect on the reliability. In figure 13(c) the
inter-failure intervals are increasing, and replacement items
have a clear tendency to be better (ie to last longer) than
the items they replace. In this case the I.I.D. assumption is
clearly invalid - later componenents come from life-time
distributions with higher mean values than earlier ones. This
behaviour is reflected in a reducing failure rate in the system.
An equally clear picture emerges from Figure 13(d) but with the
opposite effect of replacement items being worse than those they
replace, reflected in an inreasing system failure rate. Figure
13(b) shows no trend that is obvious by inspection, and in 13(a)
there is no way in which a trend could be present.

Weibull analysis would not distinguish in any way between these
obviously very different situations. In all four cases the same
result would be obtained with $\hat{\beta}$ = 1.5 and $\hat{\eta}$ = 100.

The fact that the failures in 13(c) and 13(d) are not I.I.D., and
cannot therefore be analysed by a procedure that attempts to fit
a single distribution to them, is obvious by inspection. In
practice, the situation is usually less clear, and some form of
objective test for I.I.D. is required. A suitable test is
available in the Laplace test for the trend previously described
in 4.6. This should be applied to component data originating
from field service results to justify the I.I.D. assumption
before proceeding with distribution fitting techniques such as
Weibull analysis. The test statistics for this example are:
         Fig 13(b) -   -0.15 (not significant)
         Fig 13(c) -   -1.79 (significant at 5%)
         Fig 13(d) -   1.79 (significant at 5%)
(using the expression for the failure terminated case)

43

```
[--67---x
[-----127-----x
[-32-x
[-----125-----x                    (a) Test Rig
[----96---x
[19x
[--------196--------x


167 127 32 225 9619 194    52      (b)Sequenced
[--x----xx----x--xx-------x-x         Data
                                     (No trend)


[19 52   96  125 127    194        (c)Sequenced
[xx-x--x---x---x----x-------x          Data
[3267                                (Increasing Mean)


[   194    127 125 96    52 19     (d)Sequenced
[-------x----x---x---x--x-xx           Data
[                 67 32               (Reducing Mean)
```

FIGURE 13   Failure Sequences

4.8 Component and System Burn-In

As previously indicated, engineering development in a system will
generate, in the long-term, 'better than new' replacements for
failed items, the resultant reducing failure rate being described
as 'reliability growth'. A similar effect on a much shorter
time-scale can often be observed on individual systems in a
'frozen' state of development with no attempt being made to
incorporate design improvements. This early reducing failure
rate is due to the presence in the system of 'weak' elements that
are in some way sub-standard and will come from distribution with
lower expected life-times than "typical" elements. They are
therefore expected to fail early in their lives, and be replaced
not with further "weak" items, but with "typical" ones. The
system failure rate will reduce until all such failures have
occurred.

Planned 'burn-in' of systems is the activity of operating them
(often under accelerated environments) for a suitable period
before installation to precipitate and remove these early
failures.

A similar approach can be applied to components - in a given
batch of components there may be some 'weaklings' that can be
made to fail by operating the components in some artificial test
before installation in the system. In this case the 'system' is
the batch of components and it will exhibit a reducing failure
rate as the sub-standard components fail early. This must not be

44

confused with the individual components exhibiting a reducing hazard function - the two concepts are quite separate.

The practicalities of 'burn-in' testing will be covered in Chapter 9. A comprehensive review of the subject is given by Jensen and Petersen (4).

References

1.   O'Connor, P.D.T. Practical Reliability Engineering.   1986, J. Wiley, London.

2. Nelson, L.S. Weibull Probability Paper 1963. Indust.Qual.Cont. Vol 20, 452-453.

3. Nelson, W.   Hazard Plotting for Incomplete Data, 1969, Journal Quality Technology, Vol 1, No.1, 27-52.

4.   Jensen, F and Petersen, N.E. Burn-In, 1982, J. Wiley, New York.

4

System Reliability Modeling

R. N. ALLAN
**Department of Electrical Engineering
and Electronics
UMIST
Manchester, UK**

RELIABILITY ASPECTS OF SYSTEMS

Concepts

Systems generally fall into one of two types; continuously
operated/repairable systems and "mission" orientated systems.

Continuously operated systems are expected to function
continuously in a successful state but in practice cycle between
the up state and the down state. Typical examples are electric
power supplies feeding motors, pumps, computers, instruments
etc., which are used to operate or monitor processes such as
chemical plants, nuclear reactors, etc. In the event of a system
failure, the process plant must be shut down because of either
lack of driving power or potential hazardous outcomes. One of the
most important reliability parameters for these systems is their
limiting state unavailability, i.e. the time-independent
probability of being found in the down state.
Mission orientated systems can be divided into two categories.
The first are those systems which remain idle for much of their
time but are required to function without failure when needed.
These systems are used, for example, for the safe tripping and
cooling of nuclear reactors in the event of an undesireable
initiating event such as loss of coolant, for safely shutting
down a chemical plant when a potentially hazardous occasion
arises, in the life cycle of a military missile and in the
protection system of transmission lines. In these cases the start
of the operating phase is not known and is therefore a random
event. The reliability parameter of importance is a
time-dependent probability that defines the probability of the
system being available at the start of the operating phase and of
no failures occurring during the mission. The second category are
those systems for which the start of the mission is known, which
are expected to function continuously without failure for a
specific mission period and which are not repairable during the
mission. A typical example is the electrical control system of an
aircraft. These systems can be repaired and maintained between
missions. In these cases, the important reliability parameter is
the probability of first failure of the system or the mean time
to first system failure.

In order to improve system reliability, two main features are used; redundancy and diversity.

Redundancy

Redundancy is of two forms; parallel and standby. The system shown in Figure 1 contains both forms. This represents the hypothetical station supplies of a nuclear generating station and in particular identifies a guaranteed supply point (GSP) busbar. It could also represent the power supply of chemical plants, steel mills or any other process plant.

Parallel redundancy is illustrated by the parallel branches labelled 1 and 2. This type of redundancy is frequently used provided the redundant components operate in the same way. The main advantage of this form of redundancy is that, in the event of a failure of one component or subsystem, the others continue to satisfy system demand without interruption. Its disadvantage is that the failure rate of many components is greater when operating than when idle and therefore the probability of failure may be increased.

Several instances of standby redundancy are illustrated in Figure 1. If transformer T1 fails, breaker B1 can be closed so that the supply to the GSP is changed from the 400kV supply to the 132kV supply. The normal supply to the GSP is through the rectifier/inverter branch in order to obtain a high quality

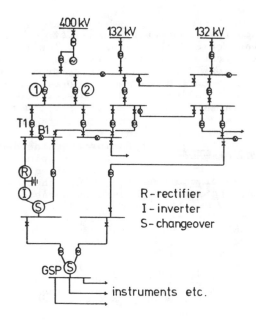

Figure 1   Typical supply system

stabilised supply. If the supply to this branch fails, thebattery supply can continue supplying the load for a short period, e.g. one half hour. If this is insufficient or the failure is in the inverter branch, a rapid changeover switch restores supply through an unstabilised bypass branch. If both of these branches fail, another changeover switch can recover the supply through a third branch fed from a separate part of the system or from standby deisel generators or gas turbines.

Diversity

Diversity is used in practical systems in order to minimise common mode (or cause) failures and to prevent one system failure event affecting all branches of a redundant system. The following examples illustrate these points. The redundant components should be separated physically to reduce the chance of fire, accidental damage or wilful damage affecting all components. Diverse components are desireable, e.g. the rectifier/inverter system and unstabilised system in Figure 1, to reduce the chance of a common manufacturing fault or adverse stress and temperature environment. The redundant branches should be fed from separate supplies as illustrated in Figure 1 to prevent one system failure affecting all branches.

Diversity can be achieved, as shown hypothetically in Figure 2, by monitoring several system parameters and connecting these diverse measurements through majority vote or m-out-of-n gates.

Evaluation Techniques

Several techniques can and have been used to evaluate system reliability. These include Monte Carlo simulation; fault trees, event trees, truth tables, Markov modelling, network reduction, minimal cut sets and conditional probability techniques. The most appropriate technique depends on the system, its operational characteristics, its modes of failure and the consequences of failure. The method should therefore reflect and respond to the

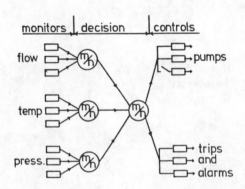

Figure 2   Simplified control system

actual system behaviour. Sections of this Chapter dealing with
these techniques have been reproduced with permission from
Reference 1.

SERIES/PARALLEL SYSTEMS

Modelling Concepts

Many systems are either physical networks or can be represented
as networks in which the components are connected either in
series, parallel, meshed or a combination of these. This section
considers only series and parallel networks. It is vital that the
relationship between the system and its network model is
thoroughly understood since the actual system and its reliability
model may not necessarily have the same topological structure.

A series system is defined as a set of components which, from a
reliability point of view, must ALL work for system success or
only ONE needs to fail for system failure. A parallel system is
defined as a set of components for which, from a reliability
point of view, only ONE needs to work for system success or ALL
must fail for system failure.

A series system therefore represents a non-redundant system and a
parallel system represents a fully redundant system.

Series Systems

Consider a series system consisting of n independent components.
The requirement for system success is that all components must be
working. From basic probability laws [1] this means that the

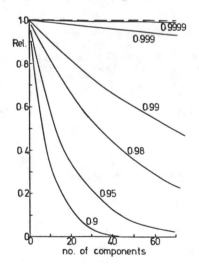

Figure 3   Effect of series components

probability of system success or reliability is:

$$R_S = \prod R_i \qquad 1.$$

where R_i = probability of successful operation (reliability) of component i

and $Q_S = 1 - R_S = 1 - \prod R_i \qquad 2.$

Because of the product rule of Equation 1, the reliability of a series system decreases as the number of components increases. This is illustrated in Figure 3.

The concept of Equation 1 applies to both time independent and time dependent probabilities. If the value of R_i obeys the exponential distribution with a failure rate of λ_i then [1]:

$$R_S(t) = \prod \exp(-\lambda_i t)$$

$$= \exp(-\Sigma \lambda_i t) \qquad 3.$$

If a single equivalent component having a failure rate λ_E is used to represent the series system then

$$R_S(t) = \exp(-\lambda_E t) \qquad 4.$$

and $\lambda_E = \Sigma \lambda_i \qquad 5.$

which indicates the equivalent failure rate of a series system whose components obey the exponential distribution is the summation of the failure rates of the individual components.

Parallel Systems

Consider a parallel system consisting of n independent components. The system requirement is that all components must fail to cause system failure. The system unreliability can therefore be obtained from basic probability laws [1] as:

$$Q_P = \prod Q_i \qquad 6.$$

where Q_i = unreliability of component i

and $R_P = 1 - Q_P = 1 - \prod Q_i \qquad 7.$

Table 1   Reliability of Parallel Systems

number of components	system reliability	incremental reliability
1	0.800000	0.000000
2	0.960000	0.160000
3	0.992000	0.032000
4	0.998400	0.006400
5	0.999680	0.001280
6	0.999936	0.000256

In the case of parallel systems, the system unreliability decreases as the number of parallel components is increased and hence the reliability increases with the number of components. This is illustrated in Table 1 for a system having components with a reliability of 0.8. This Table also shows the increase in reliability obtained by adding each additional component.

It is evident that the addition of the first redundant component to the one-component system provides the largest benefit to the system, the amount of improvement diminishing as further additions are made.

The concept of Equation 6 also applies to time dependent probabilities. Therefore if the exponential distribution is again applicable [1]:

$$Q_P(t) = \prod (1 - \exp(-\lambda_i t)) \qquad\qquad 8.$$

In this case however a single equivalent failure rate cannot be derived to represent the complete parallel system because the system unreliability can be expressed only as a series of exponential functions.

Series/Parallel Systems

The series and parallel systems discussed previously form the basis for analysing more complicated configurations. The general principle used is to sequentially reduce the complicated configuration by combining appropriate series and parallel branches of the reliability model until a single equivalent element remains. This equivalent element then represents the reliability (or unreliability) of the original configuration. This technique [1] is generally known as a (network) reduction technique.

Partially Redundant Systems

The previous sections have been concerned with only series systems (non-redundant) and parallel systems (fully redundant). Some systems may contain regions that are partially redundant or m-out-of-n. In such cases, network reduction can still be used but the series/parallel equations must be replaced [1] by state enumeration or the use of the binomial distribution. For example, consider a 2-out-of-3 system.

If all components are identical, the binomial distribution gives:

$$R_P = R^3 + 3R^2Q \quad \text{and} \quad Q_P = 3RQ^2 + Q^3$$

If all components are non-identical then state enumeration methods [1] can be used.

This gives the reliability (or unreliability) of those regions containing partial redundancy using either time independent or time dependent values. These values of R_P and Q_P are then inserted into the network model of the complete system which is then solved using the previous reduction techniques of series/parallel configurations.

Mean Time to Failure

The MTTF of a series system is given by [1]:

$$MTTF = m = \int_0^\infty R(t) \, dt \qquad 9.$$

For components which obey the exponential distribution; this MTTF is given by [1]:

$$m = 1 / \lambda_E \qquad 10a.$$

The MTTF for a parallel system is more complicated. For a 2 component parallel system [1]:

$$m = \frac{1}{\lambda_1} + \frac{1}{\lambda_2} - \frac{1}{\lambda_1 + \lambda_2} \qquad 10b.$$

Numerical Application

Consider a control system consisting of 3 main components; a transducer, an amplifier and an actuator, having failure rates of 0.35, 0.05 and 0.02 f/yr, respectively. If all components operate in their useful life period, evaluate:

(a) the probability of this system surviving for one year without failure and its MTTF.

(b) the minimum number of transducers which must be connected in parallel redundancy to increase the probability of the system surviving for one year without failure to greater than 0.9.

The solution to this example is:

(a) This is a series system, so

$\lambda_E = \Sigma \lambda = 0.35+0.05+0.02 = 0.42$ f/yr

$R(1 \text{ yr}) = \exp(-0.42 \times 1) = 0.657047$

and MTTF $= 1/0.42 = 2.38$ yr

(b) let the transducer, amplifier and actuator be components 1, 2 and 3 respectively, then

$R_2(1 \text{ yr}) = 0.951229$
$R_3(1 \text{ yr}) = 0.980199$
and $R_1 > 0.9/(R_2 \times R_3)$
 or $Q_1 < 0.034743$

with two transducers, $Q_1 = 0.087209$
with three transducers, $Q_1 = 0.025754$

i.e. a minimum of 3 transducers are required

STANDBY SYSTEMS

Perfect Switching

One additional problem in standby systems is that a failure
sensing and changeover device is required to bring the standby
unit into operation when the main component fails. These elements
are additional items not required in parallel redundant systems
and therefore can affect the overall reliability of the system.

In order to illustrate the techniques which can be used to
analyse these systems, consider the basic standby system shown in
Figure 4 in which A represents the main operating component, B
the standby component and S the sensing and changeover switch.

Assume that the switch is 100% reliable, that the standby
component does not fail while in the standby mode, that the
components are identical and that they obey the exponential
distribution with a failure rate λ.

Consider first a 2 component system having a single main
component and one standby component.

This arrangement can be regarded [1] as an equivalent single unit
which is allowed to fail once. After the first failure of the
equivalent unit (failure of A), the standby component (B) takes
over for the remainder of the mission and therefore the system
does not fail. If there is a second failure of the equivalent
unit (failure of B), the system also fails. This implies that the
Poisson distribution can be used to evaluate the probability of
system failure. Since the probability that x components fail in
time t is given by [1]:

$$P_x(t) = \frac{(\lambda t)^x e^{-\lambda t}}{x!}$$

then:

P[no components fail] = $P_0(t)$
$= e^{-\lambda t}$

P[exactly 1 component fails] = $P_1(t)$
$= \lambda t e^{-\lambda t}$

and the reliability of the system is

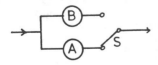

Figure 4   Simple standby system

$$R(t) = P_0(t) + P_1(t) = e^{-\lambda t}(1+\lambda t) \quad 11.$$

This can be extended to any number of standby components since the number of failures that can be tolerated is equal to the number of standby components. Therefore in the general case of n identical standby components [1]:

$$R(t) = \sum_{x=0}^{n} \frac{(\lambda t)^x e^{-\lambda t}}{x!} \quad 12.$$

i.e., the probability of system failure is given by the sum of the first n terms of the Poisson distribution.

The values of MTTF for these systems can be evaluated using the principle of Equation 9 giving for n standby components [1]:

$$m = (n + 1)/\lambda \quad 13.$$

Imperfect switching

It was assumed in the previous section that the sensing and changeover device was 100% reliable. It can be readily appreciated that this is unlikely, and that the reliability of this device is less than unity.

Define P_s as the probability of successful operation of the sensing and changeover device. The value of P_s can be established in practice from a data collection scheme which records the number of successful and failed operations of the device since

$$P_s = \frac{\text{number of successful operations}}{\text{number of requested operations}} \quad 14.$$

The most likely cause for the sensing and changeover device to fail to operate when needed is a failure of the device between the last time it was operated, tested or maintained and the occasion it is required to operate. These faults or failures are frequently termed unrevealed faults [2] since they are not apparent until a subsequent action such as operation, testing or maintenance is performed.

In the case of the 2-component standby system, system success requires that either no failures occur or one failure occurs and the switching device operates. This logic gives [1]:

$$R(t) = e^{-\lambda t}(1 + P_s \lambda t) \quad 15.$$

This concept can be extended [1] to the case of n standby components since each term other than that associated with zero failures must be weighted by the value of P_s. An additional complexity is that the value of P_s may be a function of time, therefore a variable, and different for each term of these equations.

Numerical Example

In order to illustrate the effect of P_s on the reliability of a standby system, consider a 2 component system each having a

failure rate of 0.02 f/hr and a mission time of 10hr. The results
are shown in Table 2.

Table 2   Effect of P_s

P_s	$R(t)$
1.00	0.982477
0.99	0.980839
0.98	0.979202
0.97	0.977565
0.96	0.975927
0.95	0.974290
0.94	0.972652
0.93	0.971015
0.92	0.969377
o.91	0.967740
0.90	0.966102

The reliability for a 2-component parallel system is 0.967141. It
follows therefore that the value of P_s significantly degrades the
reliability of the standby system and can cause its reliability
to be less than that of the parallel system.

FAILURE-TO-SAFETY AND MAJORITY VOTING

If a system failure occurs which creates a potentially hazardous
situation, then it is essential to ensure, as far as practical,
that the system is put into a safe condition. This concept is
known as failure-to-safety or fail-safe. The probability of
achieving this cannot be unity although it can be made very close
to unity. It leaves therefore, a very small probability of the
system remaining in a dangerous condition, this being known as
failure-to-danger or fail-danger. The system should therefore be
designed to maximise the probability of fail-safe and minimise
the probability of fail-danger. This can be achieved by
redundancy and diversity as described previously.

Consider the coolant flow detector system in Figure 2. If each
detector has a reliability of 0.99 and an unreliability of 0.01,
then the probability of at least one, two and three detectors
operating successfully is 0.999999, 0.999702 and 0.970219
respectively. These results suggest that, for the greatest
probability of fail-safe, the majority vote gate should be
1-out-of-3, i.e. a fully parallel redundant system. This analysis
however ignores the possibility that a detector can operate
inadvertently, i.e. give a false operating signal. If such a
signal was given, the system would then be shut down
unnecessarily.

Consider the same system and assume that the probability of a
detector giving a false signal is 0.01 and the probability of not
giving a false signal is 0.99. This assumption decouples the
present consideration from the previous one. Strictly the two
effects should be considered together and this decoupling is used

only to illustrate the concept. In the present consideration,
probability of three, two and one detectors giving a false signal
is 0.000001, 0.000297 and 0.029403 respectively. Consequently the
probability of shutting the system down unnecessarily with a
1-out-of-3 gate is 0.029701 and with a 2-out-of-3 gate is only
0.000298; giving a ratio of almost 100:1. For this reason,
majority vote elements are normally designed as 2-out-of-n since
this greatly reduces the probability of shutting down the system
unnecessarily without significantly affecting the probability of
failure-to-safety.

CONDITIONAL PROBABILITY APPROACH

The previous techniques can only be applied to networks having
series/parallel structures. Many systems are not as simple, e.g.
the bridge-type network shown in Figure 5. There are a number of
techniques which can be used to solve this type of network,
including the conditional probability approach and minimal cut
set analysis.

The conditional probability approach [1] involves sequentially
reducing the system into subsystem structures that are connected
in series/parallel and then to recombine these subsystems using
the conditional probability method. The conditional probability
concept is:

    P(system success or failure) = P(system success or failure
    given component X is good). P(X is good) + P(system success
    or failure given component X is bad). P(X is bad)   16.

In order to explain this approach, consider the system shown in
Figure 5 in which success requires that at least one of the
paths, AC, BD, AED, BEC is good. The best component to choose as
X is component E in the present example.

The system is now subdivided into two subsystems, one with E
considered good, i.e. it cannot fail and one with E considered
bad, i.e. it is always failed. This subdivision is shown in
Figure 6.

Visual inspection indicates that the two subsystems are simple
series/parallel structures. In some systems, one or more of the
subsystems may need further subdivision before a series/parallel
structure is obtained.

In the present example, the overall system reliability will be:

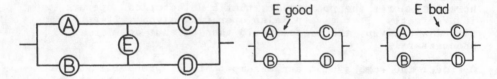

Figure 5 Bridge network   Figure 6   Conditional probability
                                        approach

$R_S = R_S$(if E is good) $R_E + R_S$(if E is bad) Q_E

where

R_S(if E is good) $= (1-Q_A Q_B)(1-Q_C Q_D)$

R_S(if E is bad) $= 1- (1-R_A R_C)(1-R_B R_D)$

If all components are identical:

$R_S = 2R^2 + 2R^3 - 5R^4 + 2R^5$ 17.

If the reliability of each component is 0.99, then $R_S = 0.99979805$ and $Q_S = 0.00020195$.

MINIMAL CUT SET METHOD

The minimal cut set method [1] is a powerful tool for evaluating the reliability of a system and is the basis for most of the evaluation methods for distribution, communication and other flow-type networks. It is defined as:

"A minimal cut set is a set of system components which, when failed, causes failure of the system but when any one component of the set has not failed, does not cause system failure."

This definition means that ALL components of a minimal cut set must be in the failure state to cause system failure. If the evaluation is carried out in its most complete form, the result is exact and no approximations are made. Simplifications are made in practice however which generally introduce negligible errors.

Four minimal cut sets of the system shown in Figure 5 can be identified; these being

(AB) (CD) (AED) (BEC).

From the definition of minimal cut sets it is evident that the components of each cut set are effectively connected in parallel and the failure probabilities of the components in the cut set may be combined using the principle of parallel systems. The unreliability of the system is then given by [1]:

$Q_S = P(C1 \cup C2 \cup C3 \cup C4)$
   $= P(C1) + P(C2) + P(C3) + P(C4) - P(C1 \cap C2) - P(C1 \cap C3) - P(C1 \cap C4) - P(C2 \cap C3) - P(C2 \cap C4) - P(C3 \cap C4) + P(C1 \cap C2 \cap C3) + P(C1 \cap C2 \cap C4) + P(C1 \cap C3 \cap C4) + P(C2 \cap C3 \cap C4) - P(C1 \cap C2 \cap C3 \cap C4)$ 18.

This equation is exact and if all components are identical, becomes:

$Q_S = 2Q^2 + 2Q^3 - 5Q^4 + 2Q^5$

from which, if R = 0.99 and Q = 0.01, then $Q_S = 0.00020195$ and $R_S = 0.99979805$. These are identical to those given by the conditional probability approach. In practice, the unavailability is usually very small and approximations can be made. This involves neglecting all terms except the summation of individual cut set probabilities, i.e.

$$Q_S = P(C1) + P(C2) + P(C3) + P(C4) \qquad 19.$$

This approximation gives the upper bound to system unreliability and introduces negligible errors for most applications. With this approximation:

$$Q_S = Q_A Q_B + Q_C Q_D + Q_A Q_D Q_E + Q_B Q_C Q_E \qquad 20.$$

or if all components are identical:

$$Q_S = 2Q^2 + 2Q^3$$

In this case, if $R = 0.99$ and $Q = 0.01$, then $Q_S = 0.000202$ and $R_S = 0.999798$.

A second approximation can be made by neglecting high order cuts. If the two second order cuts only are considered in the present example, then:

$$Q_S = Q_A Q_B + Q_C Q_D$$

or if all components are identical, $Q_S = 2Q^2$ which gives for $R = 0.99$ and $Q = 0.01$, $Q_S = 0.000200$ and $R_S = 0.999800$.

It is evident that these approximations introduce very small and generally ignorable errors.

MULTI FAILURE MODES

Previously it was assumed that only one failure mode exists. If there are several modes but all have the same effect on the system, they can be grouped together and the previous analysis is applicable. In some cases their effect is different and they can not be grouped. An example is open circuits and short circuits of electrical networks. The previous techniques can be extended [1] to include such effects. This is also useful to demonstrate the application of the conditional probability approach. Consider the diode problem shown in Figure 7 and assume both diodes are identical.

The conditional probability approach gives [1]:

$R_S = P(\text{system success} \mid 1 \text{ normal}).P(1 \text{ normal}) + P(\text{system success} \mid 1 \text{ open}).P(1 \text{ open}) + P(\text{system success} \mid 1 \text{ shorted}).P(1 \text{ shorted}).$

Let P_N = prob (normal operation)
    P_O = prob (open circuit)
    P_S = prob (short circuit), then

given 1 normal: P(system success) = P_N + P_0
given 1 open: P(system success) = P_N
given 1 shorted: P(system success) = 0
and

$$R_S = P_N^2 + 2P_N P_0$$

Consider the following numerical examples:

If P_N = 0.98 and P_0 = P_S = 0.01 then R_S = 0.98
If P_N = 0.98, P_0 = 0 and P_S = 0.02 then R_S = 0.9604
If P_N = 0.98, P_0 = 0.02 and P_S = 0 then R_S = 0.9996.

FAULT TREES

The fault tree method has been used widely for many years for
assessing the reliability of standby and mission orientated
systems. It is very rarely used for topological types of systems.
In this method a particular failure condition is considered and a
tree is constructed that identifies the various combinations and
sequences of other failures that lead to the failure being
considered.

This method is frequently used as a qualitative evaluation method
in order to assist the designer, planner or operator in deciding
how a system may fail and what remedies may be used to overcome
the causes of failure. The method can also be used for
quantitative evaluation, in which case the causes of system
failure are gradually broken down into an increasing number of
hierarchical levels until a level is reached at which reliability
data is sufficient or precise enough for a quantitative
assessment to be made. The appropriate data is then inserted into
the tree at this hierarchical level and combined together using
the logic of the tree to give the reliability assessment of the
complete system being studied.

In order to illustrate the application of this method, consider
the electric power requirements of a system. In this example, the
failure event being considered is 'failure of the electric
power'. In practice the electric power requirements may be both
a.c. power, to supply energy for prime movers, and d.c. power, to
operate relays and contactors, both of which are required to
ensure successful operation of the electric power. Consequently
the event 'failure of electric power' can be divided into two
subevents 'failure of a.c. power' and 'failure of d.c. power'.

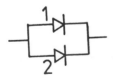

Figure 7   Two parallel diodes

This is shown in Figure 8 with the events being joined by an OR gate as failure of either, or both, causes the system to fail.

If this subdivision is insufficient, both subevents must be divided further. The event 'failure of a.c. power' may be caused by 'failure of offsite power', i.e., the grid supply or by 'failure of onsite power', i.e., standby generators or similar devices. The fault tree of Figure 8 can be subdivided as shown in Figure 8 with the events joined by an AND gate since both must fail in order to lose the a.c. power. This process can be continued downwards to any required level of subdivision. The logic used requires a thorough engineering understanding of the system being analysed since it is necessary to appreciate the failure events that can occur as well as how they are linked together. The linking process may involve AND and OR gates as shown in Figure 8 together with other logic gates such as NOT, NOR, NAND and m-out-of-n gates, i.e., majority vote components.

After developing a fault tree, it is necessary to evaluate the probability of occurrence of the upper event by combining component probabilities using basic rules of probability and the logic defined in the fault tree [1].

In the present example associated with Figure 8, let the reliability of the offsite power (R_1) be 0.933, the reliability of the onsite power (R_2) be 0.925 and the reliability of the d.c. power (R_3) be 0.995, then

$$
\begin{aligned}
Q(\text{a.c. power}) &= (1-R_1)(1-R_2) \\
&= (1-0.933)(1-0.925) \\
&= 0.005025 \\
R(\text{electric power}) &= R(\text{a.c. power})R_3 \\
&= (1-0.005025)\,0.995 \\
&= 0.990000
\end{aligned}
$$

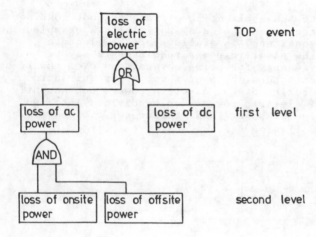

Figure 8   Fault tree

Q(electric power) = 0.010000

EVENT TREES

Concepts

An event tree simulates, not only the topology of a system, but more importantly, its sequential operational logic. Consequently it is an ideal method for evaluating the reliability of systems which involve the sequential operation of control circuits and devices, interlocks, standby plant and opening and closing of switches and circuit breakers.

An event tree is constructed [1] by commencing with the initiating event which may be an unsatisfactory operating event such as a system fault or may be a demand for a system to change from one operating state to another. The sequence of events that may then occur or not occur are constructed in chronological order. This produces a rapidly expanding tree-structure which is terminated when all the event sequences have been included. Finally, the outcome of the system due to each path of the tree must be ascertained from an engineering understanding of the system behaviour.

Many studies have been conducted at UMIST into the application of event trees. These have included:

(a) Pumped storage hydro generation. Pumped storage stations may reside in one of several modes; stand-still, pumping, generating. Operationally, it may be required to transit rapidly from one mode to another particularly under emergencies. This involves operation of control circuits, interlocks, circuit breakers as well as the turbines themselves. These operational sequences can be analysed [3] using event trees to give the probability of successful response to the operational request.

(b) Emergency standby generation. When a normal supply system fails, emergency generating plant may need to be operated. This can occur in conventional and nuclear stations, in industrial process plants, in hospitals. This involves the sequential operation of control systems, interlocks, opening and closing of breakers, start up of the standby generators. This can again be analysed [4,5] using event trees.

(c) Protection systems. A protection scheme requires the circuit breakers to operate following the sequential operation of

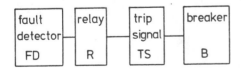

Figure 9   Simplified protection system

61

fault detectors and relays. An example is given below. The technique has been used [6] to quantify station originated outages in composite system reliability evaluation.

(d) <u>Nuclear power stations</u>. One important aspect of nuclear stations is safety and the successful shutdown of the station following an abnormal event. This involves the sequential operation of emergency safety devices which can be analysed [7-10] using the event tree method.

System and Event Tree

In order to illustrate the conceptual aspects of event trees, consider the very generalised form of a protection system shown in Figure 9. Consider also a particular component that is protected by two breakers B1 and B2. Assume that the same fault detector(FD), same relay (R) and separate trip signal devices, TS1 and TS2, actuate the breakers B1 and B2 respectively. This is a simple example and more redundancy, diversity and independence will generally be included in a practical system. It does however illustrate the modelling and evaluation procedure. The event tree for this system is shown in Figure 10. This illustrates the sequence of events together with the outcomes for each event path, some of which lead to both breakers operating successfully but others lead to both breakers failing simultaneously.

This event tree assumes that each component can reside in one of two states. If required, other states can be included by creating three or more branches at each node and deducing thr appropriate system outcomes for each of the new paths.

Evalulating Event Probabilities

The event probabilities are the probabilities that each device will and will not operate when required. These are time dependent probabilities. The time at which a device is required to operate however is a random variable and unknown in advance. Two

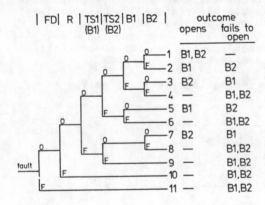

Figure 10   Event tree

particular indices can be evaluated for these events.

(a) <u>Time dependent values</u>. The probability of failure of a component or device can usually be characterised [1] by a distribution such as the exponential, Weibull, lognormal, etc. If the parameters of the distribution are known, the value of failure probability can be evaluated at any time in the future. This can acccount [7] for inspection, checks, replacement, etc. Using these time dependent values, the event probabilities and therefore the system probabilities can be evaluated as a function of time. Also the effect of inspection intervals, replacement policies, etc can be evaluated and optimum schedules established.

(b) <u>Average unavailability</u>. An alternative index is the average behaviour of the device between consecutive inspection tests measured in terms of average unavailability.

If the time between consecutive tests is T_c, then the average unavailability U is:

$$U = \frac{1}{T_c} \int_0^{T_c} Q(t) \, dt \hspace{2cm} 21.$$

where $Q(t)$ is the time dependent failure probability of a component. In the case of the exponential distribution:

$$U = 1 - \frac{1}{\lambda T_c} (1 - e^{-\lambda T_c})$$

$$\approx \frac{\lambda T_c}{2} \quad \text{if } \lambda T_c \ll 1 \hspace{2cm} 22.$$

A similar evaluation can be applied for other distributions using, if necessary, a numerical integration method.

Evaluating Outcome Probabilities

The outcome probabilities are readily evaluated after the event tree and event probabilities have been deduced. First the event paths leading to each outcome are identified. The probability of occurrence of each path is the product of the event probabilities in the path. The outcome probability is then the sum of the probabilities of each path leading to that outcome.

Table 3   Outcome probabilities for Figure 10

outcome	paths	probability
B1 & B2 operate	1	0.970094
B1 op's, B2 fails	2,5	0.009289
B1 fails, B2 op's	3,7	0.009289
B1 & B2 fail	4,6,8-11	0.011328

Consider the situation in the present example when the failure rates are 4×10^{-2} f/yr for the fault detector, 5×10^{-3} f/yr for the relay, 3×10^{-2} f/yr for each trip signal device and 8×10^{-3} f/yr for each breaker. Also assume an inspection interval of 6 months. Using this data and Equation 22 gives the outcome probabilities shown in Table 3.

Tie Sets and Minimal Cut-Sets

The paths associated with the event tree are tie sets and can be used directly to deduce the minimal cut sets of the system outcomes. The details of the deduction process are described [1,11] elsewhere. In brief they can be evaluated either from the paths leading to the system outcome under consideration or from the paths NOT leading to that system outcome.

Consider the paths (tie sets) leading to the system outcome. These paths contain the components that must fail or the events that must occur together in order to cause the system outcome. The cut sets can therefore be deduced by enumerating the components that are failed in each path. These cut sets may not be minimal and therefore they must be searched to reduce them to a minimal set.

Consider the system outcome of (B1 and B2 failing to operate). The paths of Figure 10 leading to this outcome are 4, 6, 8-11. The minimal cut sets identified in these paths are (B1,B2), (TS1,B2), (TS2,B1), (TS1,TS2), (R), (FD). In this example, these can be deduced by inspection but in more complicated systems, the formal approach [11] is needed to identify the minimal cut sets.

CONTINUOUS MARKOV PROCESSES

General Modelling Concepts

Most reliability evaluation problems are normally concerned with systems that are discrete in space and continuous in time, i.e. they exist continuously in one of the system states until a transition occurs which takes them discretely to another state. The techniques described in this section pertain to stationary Markov processes i.e. the system is memory-less and failure and repair characteristics of the components obey exponential distributions (see Reference 1).

Consider first the case of a single repairable component having a

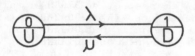

Figure 11   Single component diagram

failure rate λ and repair rate μ. The state transition diagram for this component is shown in Figure 11.

The most rigorous method for solving this and similar processes is to set up [1] a differential equation by considering the probability of occurrence of events in a small incremental interval of time dt. This method proves intractable for anything but the smallest of systems and alternative numerical techniques are generally used. These alternative methods do not establish generalised expressions representing state probabilities but, instead, produce direct numerical results. In order to illustrate the generalised solutions given by a differential equation method, consider the model shown in Figure 11. If the system starts in the UP state, the solution for this can be shown [1] to be:

$$P_0(t) = \frac{\mu}{\lambda+\mu} + \frac{\lambda}{\lambda+\mu} e^{-(\lambda+\mu)t} \qquad 23.$$

$$P_1(t) = \frac{\lambda}{\lambda+\mu} - \frac{\lambda}{\lambda+\mu} e^{-(\lambda+\mu)t} \qquad 24.$$

where $P_i(t)$ is the probability of residing in state i at time t, and $P_0(0)=1$, $P_1(0)=0$.

These values of $P_i(t)$ are known as the time dependent state probabilities. If the system is ergodic, i.e. all states communicate directly or indirectly with all others, then $P_i(t)$ will tend to limiting or steady state values as $t \to \infty$. If these are defined as P_0 and P_1 then:

$$P_0 = \mu/(\lambda+\mu) \qquad 25.$$

$$P_1 = \lambda/(\lambda+\mu) \qquad 26.$$

For the exponential distribution:

$$MTTF = m = 1/\lambda \quad \text{and} \quad MTTR = r = 1/\mu \qquad 27.$$

Substituting into Equations 25 and 26 gives:

$$P_0 = m/(m+r) \quad \text{and} \quad P_1 = r/(m+r) \qquad 28.$$

The values of P_0 and P_1, unlike $P_0(t)$ and $P_1(t)$, are the same irrespective of the initial state of the system. These values are known as the steady state or limiting availability A and unavailability U of the system respectively whereas Equations 23 and 24 represent the time dependent availability A(t) and unavailability U(t) of the system respectively. This value of A(t) is the probability of *being found* in the operating state at some time t in the future. This is quite different from the survivor function R(t) which represents the probability of *staying in* the operating state as a function of time. The relationship between A(t) and R(t) is shown in Figure 12.

Since the values of P_0 and P_1 are related only to mean times (Equation 28), these values are not dependent on the probability distributions associated with the components and are correct for all distributions. This is not the case however for the time dependent values $P_0(t)$ and $P_1(t)$.

Evaluating Time Dependent Probabilities

The alternative method to the differential equation technique is based on matrix multiplication [1]. In this method a state space diagram is first constructed. This is shown in Figure 11 for a single component and in Figure 13 for a two component system. A stochastic transitional probability matrix is then created. This matrix represents the transitional probabilities between states of the stochastic process. This is shown in Equation 29 below for the single component system in which element P_{ij} represents the probability of making a transition to state j during a time interval dt given that it was in state i at the beginning of the time interval. The value of dt must be chosen so that the probability of two or more transitions occurring in this interval is negligible.

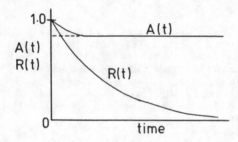

Figure 12   Availability and reliability

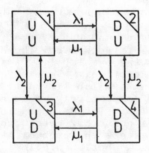

Figure 13   Two component system

66

$$P = \begin{array}{c} \text{from} \\ \text{state} \end{array} \begin{array}{c} 1 \\ 2 \end{array} \xrightarrow{\text{to state}} \begin{array}{cc} 1 & 2 \\ \begin{bmatrix} 1 - \lambda dt & \lambda dt \\ \mu dt & 1 - \mu dt \end{bmatrix} \end{array} \qquad 29.$$

The state probabilities P(t) can be evaluated at any time t into the future from:

$$P(t) = P(0)\, P^n \qquad\qquad 30.$$

where P(0) represents the state probabilities at t=0 and n=t/dt. If the system is known to be operating at t=0, then for the single component system P(0)= [1 0].

Evaluating Limiting State Probabilities

The limiting state probabilities can be evaluated [1] by repeating the matrix multiplication process of the previous section until the values of P(t) remain unchanged. This means that:

$$P(t).P = P(t) \qquad\qquad 31.$$

In the case of the single component this translates into:

$$[P_0 \quad P_1] \begin{bmatrix} 1 - \lambda dt & \lambda dt \\ \mu dt & 1 - \mu dt \end{bmatrix} = [P_0 \quad P_1] \quad 32.$$

Using Equation 32 together with the fact that $P_0 + P_1 = 1$, gives a set of equations identical to Equations 25 and 26. This evaluation would also show that the values of dt disappear in the solution and therefore, if only limiting state probabilities are required, the values of dt need not be included in the stochastic transitional probability matrix and the solution of P_i can be deduced directly from the set of simultaneous equations given by Equation 32 using numerical methods.

Evaluating System Availabilities

The values of A and U for the single component system are given directly by P_0 and P_1. For more complex systems, appropriate states must be combined. For instance, in the case of the two component system shown in Figure 13, the values of A and U depend on whether it is a series or a parallel system, i.e.

series: $A = P_1$ $U = P_2 + P_3 + P_4$ 33.

parallel: $A = P_1 + P_2 + P_3$ $U = P_4$ 34.

FREQUENCY AND DURATION TECHNIQUE

Concepts

The previous techniques permit the state probabilities and system availability and unavailability to be evaluated. It is however sometimes beneficial to evaluate [1] additional reliability indices such as the frequency of encountering a state and the average duration of residing in the state.

In order to illustrate this evaluation, first consider the single component system shown in Figure 11. In addition to the previous parameters, let T be the cycle time(=m+r) and hence f=1/T be the cycle frequency. Then, from Equation 28:

$$P_0 = \frac{m}{m + r} = \frac{m}{T} = \frac{1}{\lambda T} = \frac{f}{\lambda} \qquad 35.$$

which gives:

$$f = P_0 \lambda \qquad \text{and} \qquad m = P_0 / f \qquad 36.$$

Although this is derived for a single component system, the concept applies to all systems and, in words, is expressed as:

frequency of encountering a state
= (probability of being in the state) x (rate of departure
from the state)   37.

This concept only applies to the long term or average behaviour of the system and is not valid for time dependent frequencies. Equation 36 also shows that the mean duration of residing in a state is equal to the probability of residing in the state divided by the frequency of encountering the state.

Evaluating Individual State Indices

The basic frequency and duration concepts can be applied to any size of system. Consider a two component system, the state space diagram of which is shown in Figure 13.

The first step is to evaluate the individual limiting state probabilities. These can be evaluated using the previous techniques or, in this case, found from combining the individual component availabilities and unavailabilities. These state probabilities are shown in Table 4.

Table 4 State Indices

state	probability	frequency
1	$\mu_1 \mu_2 / D$	$\mu_1 \mu_2 (\lambda_1 + \lambda_2) / D$
2	$\lambda_1 \mu_2 / D$	$\lambda_1 \mu_2 (\mu_1 + \lambda_2) / D$
3	$\mu_1 \lambda_2 / D$	$\mu_1 \lambda_2 (\lambda_1 + \mu_2) / D$
4	$\lambda_1 \lambda_2 / D$	$\lambda_1 \lambda_2 (\mu_1 + \mu_2) / D$

where $D = (\lambda_1 + \mu_1) (\lambda_2 + \mu_2)$

The second step is to evaluate the frequency of encountering the individual states. These are obtained by multiplying the state probabilities by the rates of departure and are also shown in Table 4.

The final step is to evaluate the mean duration of each state. These are found either by reciprocating the rates of departure or using the concept of Equation 36.

Evaluating Cumulated State Indices

The individual state indices are only a partial answer and states leading to the same system outcome should be cumulated. The cumulative probability of residing in a system state is evaluated [1] by summating the state probabilities of those states which create the system state. This was done previously in Equations 33 and 34.

The cumulative frequency however must include the frequencies of all transitions that leave and enter the combined state but must ignore all transition frequencies that occur between the cumulated states since these do not represent transitions out of the combined state. Therefore, in the case of cumulated states 3 and 4 of the two component system shown in Figure 13 and Table 4:

$$f_{34} = f_3 + f_4 - \text{(frequency of encounters between 3 and 4)}$$

$$= f_3 + f_4 - (P_3 \lambda_1 + P_4 \mu_1) \qquad 38.$$

Alternatively the frequency of encountering the cumulated state can be obtained [1] by considering the expected number of transitions across the boundary wall surrounding the cumulated state. In the case of the two component system, this gives:

$$f_{34} = P_3 \mu_2 + P_4 \mu_2 \qquad 39.$$

which can be shown to be the same as Equation 38.

The final step is to evaluate the mean duration of residing in each of the cumulated system states. These values can be evaluated using the principle of Equation 36 since:

$$m_c = \frac{\text{cum prob of being in state i}}{\text{cum freq of encountering state i}} \qquad 40.$$

APPLICATION TO SPARING CONCEPTS

It has been assumed that each component is associated with two transition rates, the failure rate λ and the repair rate μ (= reciprocal of repair time r). In many cases, particularly when spares are available, three transition rates should be considered [1]; the previous two plus the installation rate γ (= reciprocal of the installation time). The cases of sparing shown in Figures 14-17 illustrate the construction of appropriate state space diagrams.

To construct [1] these state space diagrams, it is evident that,

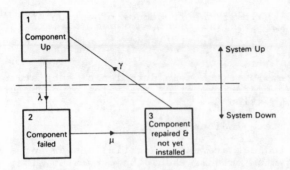

Figure 14   Single component - no spare

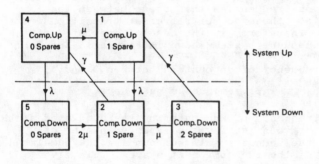

Figure 15   Single component - one spare

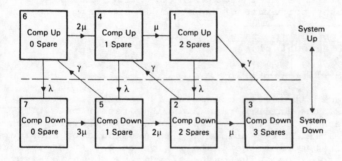

Figure 16   Single component - two spares

70

as the number of operating components increases, the number of
rows in the diagram increases and, as the number of spares
increases, the number of columns increase. A stochastic
transitional probability matrix can be created from these
diagrams and the techniques discussed previously used to find the
state and system probabilities, frequencies and durations. To
determine the economic number of spares that should be kept, it
is necessary to evaluate the availability or unavailability of
the system as a function of the number of spares. As the number
of spares increases, the incremental increase in availability
decreases, as shown in Figure 18.

The cost of keeping a large number of spares cannot be justified
if the incremental increase in availability is negligible. To
determine this limiting value, the repair action can be ignored
since there is always a spare available. The system reverts [1]
to a two state system where limiting unavailability = $\lambda_E / (\lambda_E + \gamma)$
and λ_E = appropriate system failure rate.

APPROXIMATE SYSTEM RELIABILITY EVALUATION

Concepts

Although Markov techniques form sound modelling and evaluation
methods, they become less amenable for large systems. Alternative
methods [1], based on the Markov approach, use a set of
appropriate but approximate equations. These equations relate to
a series system in which all components must operate for system
success and to a parallel system in which only one component need
work for system success. These equations can then be used in

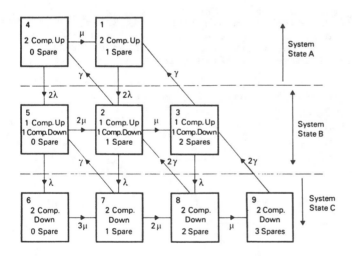

Figure 17   Two components - one spare

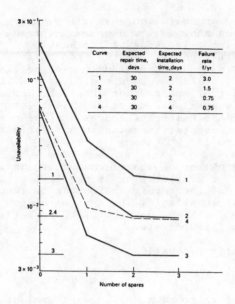

Figure 18   Effect of number of spares

conjunction with a network reduction technique or the minimal cut set approach.

Series System

Consider the case of two components connected in series, the state space diagram for which is shown in Figure 13. It is necessary to find the failure and repair rates, λ_s and μ_s, of a single component that is equivalent to the two components in series [1]. This is illustrated in Figure 19.

The probability of the single equivalent component and the system being in the up state must be equal and given by:

$$P_{up} = \frac{\mu_1 \mu_2}{(\lambda_1 + \mu_1)(\lambda_2 + \mu_2)} \qquad 41.$$

Also the transition rate from the system up state for the single equivalent component and for the system must be equal and thus:

$$\lambda_s = \lambda_1 + \lambda_2 \qquad 42.$$

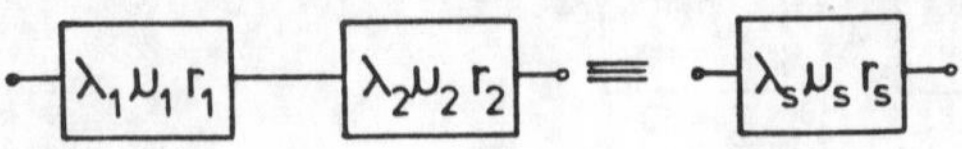

$$\boxed{\lambda_1 \mu_1 r_1} - \boxed{\lambda_2 \mu_2 r_2} \equiv \boxed{\lambda_s \mu_s r_s}$$

Figure 19   Series system

72

Substituting Equation 42 into 41 and replacing the repair rates, μ_i, by the reciprocal of the average repair times, r_i, gives [1]:

$$r_s \simeq \frac{\lambda_1 \ r_1 + \lambda_2 \ r_2}{\lambda_1 + \lambda_2} \qquad\qquad 43.$$

Finally:

$$U_s \simeq \lambda_s r_s = \lambda_1 r_1 + \lambda_2 r_2 \qquad\qquad 44.$$

Using the above logic, the failure rate, average outage duration and annual outage time of a general n-component series system is [1]:

$$\lambda_s = \Sigma \lambda_i \qquad\qquad 45.$$

$$r_s = \frac{\Sigma \lambda_i \ r_i}{\lambda_s} \qquad\qquad 46.$$

$$U_s = \Sigma \lambda_i \ r_i \qquad\qquad 47.$$

Parallel Systems

Consider the case of a two component system but in which the components are in parallel. In this case the failure rate λ_P and repair rate μ_P of a single component that is equivalent to the two components in parallel is required [1]. This is illustrated in Figure 20.

The probability of the single component and the system being in the down state must be equal and given by [1]:

$$P_{dn} = \frac{\lambda_1 \lambda_2}{(\lambda_1 + \mu_1)(\lambda_2 + \mu_2)} \qquad\qquad 48.$$

The transition rate from the down state of the two component system and the equivalent component must be equal and thus:

$$\mu_P = \mu_1 + \mu_2 \qquad\qquad 49.$$

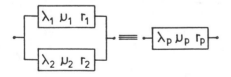

Figure 20   Parallel system

Substituting Equation 49 into 48 and replacing repair rates by the reciprocal of repair times gives [1]:

$$\lambda_P \simeq \lambda_1 \lambda_2 (r_1 + r_2) \qquad\qquad 50.$$

$$r_P = r_1 r_2 / (r_1 + r_2) \qquad\qquad 51.$$

$$U_P \propto \lambda_1 \lambda_2 r_1 r_2 \qquad\qquad 52.$$

Unlike the case of series systems, it is not possible to easily extend the equations for a 2-component parallel system to a general n-component system.

It is possible in certain systems to combine two components at a time using Equations 50 to 52. This method must be treated with utmost care because it becomes invalid if the concepts of a single failure rate per component or a single environmental state is extended to more complex situations. It is better to use an appropriate set of equations for the number of components that require combining. These equations can be deduced using the above concepts.

Equations 50 to 52 enable the rate and duration of overlapping failure events (forced outages) to be evaluated. A set of equations of similar form can be deduced [1] that evaluate the rate and duration of an event due to a forced outage overlapping a maintenance outage.

System Evaluation Techniques

Most systems do not consist of only series or parallel configurations but more often a combination of both. One method for solving these networks is to sequentially reduce the network using the equations for series and parallel combinations until the network is reduced to a single equivalent component. This method, known as network reduction, was described previously and the reliability parameters of the equivalent component are the parameters of the complete system.

An alternative method is to use the minimal cut set approach. In this approach the reliability model consists of a number of minimal cut sets connected in series and each cut set consists of a number of components connected in parallel. Therefore the equations for parallel systems can be applied to each cut set to give equivalent indices for each cut set which are then combined using the equations for series systems to give the overall system reliability indices.

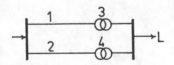

Figure 21   Dual transformer feeder

74

Numerical Example

Consider the network (dual transformer feeder system) shown in Figure 21 and the reliability data shown in Table 5.

Table 5   Reliability Data

component	λ(f/yr)	r (hr)
1,2	0.5	10
3,4	0.01	100

This system could be analysed using directly Markov modelling techniques. Alternatively, the series and parallel equations can be used as follows:

(a) using network reduction
    combining elements 1 and 3 in series gives:

λ_{13} =0.51f/yr, r_{13} =11.76hr, U_{13} =6 hr/yr

the indices of components 2 and 4 combined will be identical.

The indices for the load point are:

λ_{PP} = 6.984 x 10-4 f/yr, r_{PP} = 5.88 hr, U_{PP} = 4.106 x 10-3 hr/yr

(b) using failure mode (minimal cut set) analysis gives the results shown in Table 6.

Table 6   Reliability Results

events	λ_{PP} f/yr	r_{PP} hr	U_{PP} hr/yr
1 & 2	5.708x10-4	5.0	2.854x10-3
1 & 4	6.279x10-5	9.09	5.708x10-4
2 & 3	6.279x10-5	9.09	5.708x10-4
3 & 4	2.283x10-6	50.0	1.142x10-4
total	6.986x10-4	5.88	4.110x10-3

The second method gives much more information. It clearly indicates that the failure rate and unavailability is mainly due to the overlapping failures of the two lines, but that the average outage duration is mainly due to the overlapping outages of the two transformers. This information, vital in assessing critical areas and deducing the areas requiring investment, is not given by the network reduction technique.

COMMON MODE FAILURES

Concepts

One significant assumption that has frequently been made in
reliability evaluation, including the previous sections, is that
the failure of one component is quite independent of the failure
of any other component, both directly or indirectly. In practice
however, it has been found that a system can fail more frequently
than the predicted value by a factor of one, two or more orders
of magnitude. This is due to dependencies between failures of
components. Two important problems exist; recognition of the
modes and effects of dependence and appropriate reliability
modelling and evaluation techniques.

One of the most important modes of failure that can lead to very
high values of failure probabilities is common mode or cause
failures. This type of failure involves the simultaneous outage
of two or more components due to a common cause.

These effects have been receiving considerable attention in
recent years, e.g. References [12,13].

Many definitions have been proposed for common mode failures
although no general acceptance has yet been approved particularly
across the boundaries of the engineering disciplines. In
principle all definitions are of the form:

" a common mode failure is an event having a single external
cause with multiple failure effects which are not consequences of
one another".

There are many examples that can be given to illustrate common
mode failures and their effects. The following two are typical in
two entirely different operating situations:
a) a single fire in a nuclear reactor plant causes the failure of
both the normal cooling water system and the emergency cooling
water system because the pumps of both systems are housed in the
same pumping room
b) the crash of a light aircraft causes the failure of a two
circuit transmission line because both lines are on the same
towers.

In both of these examples, the cause is a single event, the
outage involves more than one component and neither outage is a
consequence of the other.

Other examples have been suggested as causes of common mode
failures. These are often not genuine common mode ones although
they certainly are dependent failures. These include common
manufacturer, common environment, common repair team, etc. Such
common factors can enhance component failure rates and therefore
the likelihood of a multiple failure. They must also be taken
into account.

Modelling and Evaluation

In order to include the effect of common mode failures in the
analysis of a system, it is important to be able to recognise how
they can occur and to create a suitable model.

One method that can be used, particularly for repairable systems,
is Markov modelling and analysis. The first requirement is
therefore a state space diagram that represents the behaviour of
the system. Several models have been proposed [1], one of which
is shown in Figure 22 for a two component parallel system or a
second order minimal cut set. This includes a common mode failure
rate, λ_{12}.

Such models, or adaptation and extensions of their principles,
can be included either as part of a network reduction technique
or as part of a failure modes and effects analysis. The solution
proceeds in the same way as the Markov techniques discussed in
previous sections.

Numerical Examples

In order to illustrate the effect that common mode failures have
on the availability of repairable systems, consider a simple two
identical component system having the state space diagram shown
in Figure 22. The following data was used:
component independent failure rate λ - 0.01 and 1 f/yr
component common mode failure rate λ_{12} - 0 to 10% of λ
component repair rate μ - 12, 50 and 300 repairs per yr
(equivalent to repair times of approximately 1 month, 1 week and
1 day respectively).

The variation in system unavailability is shown in Figure 23
which clearly indicates the very significant effect that only a
small percentage of common mode failures have on system
behaviour, particularly when the failure rate and repair times
are small.

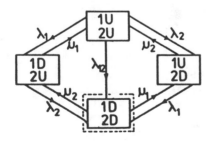

Figure 22   Common mode failure model

MONTE CARLO SIMULATION

Evaluation techniques can be divided into two main types;
analytical and simulation, the most common of which is known as
Monte Carlo simulation. The previous sections of this chapter
have been devoted to analytical techniques. These represent the
system by mathematical models and equations and evaluate the
reliability indices from these models using mathematical
solutions. Monte Carlo simulation methods, however, estimate the
reliability indices by simulating the actual process and random
behaviour of the system.

Monte Carlo simulation techniques have been used for a very wide
range of studies including load-strength [14] and power system
[15] analyses and are described in detail in Reference 16. The
basic concept of the technique is that it is a combined modelling
and sampling process. The model is a description of the system
structure and the way it behaves for a given set of input
parameters. The sampling process involves randomly selecting a
set of input parameters from the appropriate probability
distributions describing the way the input data varies. Each set
of input parameters are processed in the system model to give the
appropriate output indices. If sufficient sets of input data are
processed, the output is also described by a probability
distribution.

There are several advantages of this technique including:
a) any system structure and behaviour can be solved which may not
be possible using an analytical technique
b) the analysis can give a full statistical picture of the output
indices whereas many analytical techniques give only expected

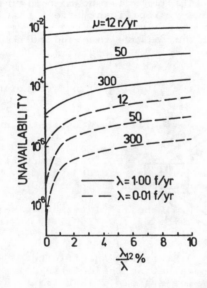

Figure 23   Effect of common mode failures

78

values and the underlying distribution remains unknown.

The main disadvantage is that it can require very large amounts of computing time before satisfactory and sufficiently accurate results are obtained.

CONCLUSIONS

Two fundamental points should be noted before applying these or similar techniques. The first is that a thorough understanding of the system, the way it operates, the way it fails and the consequences of failure should first be established. Secondly, the evaluation technique chosen should reflect and respond to the engineering understanding of the system. Misapplication of the techniques or a misunderstanding of the system behaviour can lead to conclusions that are misleading and irrelevant.

REFERENCES

1. R.Billinton, R.N.Allan, "Reliability Evaluation of Engineering Systems - Concepts and Techniques". Longmans (formerly Pitmans), 1983.

2. A.E.Green, A.J.Bourne, "Reliability Technology". Wiley, 1972.

3. D.Dalabeih, "Reliability assessment of the control system of a pumped storage scheme". MSc Dissertation, UMIST, 1978.

4. I.L.Rondiris, "Reliability evaluation of auxiliary supply systems for nuclear power stations". MSc Dissertation, UMIST, 1976.

5. K.S.Ang, "Reliability evaluation of emergency power systems". MSc Dissertation, UMIST, 1980.

6. R.N.Allan, A.Adraktas, "Terminal effects and protection system failures in composite system reliability evaluation". IEEE Trans, PAS-101, 1982, pp.4557-4562.

7. I.L.Rondiris, "Reliability evaluation of nuclear plants". PhD Thesis, UMIST, 1978.

8. A.Adraktas, "Interactive reliability evaluation of nuclear power plants". PhD Thesis, UMIST, 1980.

9. R.N.Allan, I.L.Rondiris, D.M.Fryer, C.Tye, "Computational development of event trees in nuclear reactor systems". 2nd National Reliability Conf, Birmingham 1979, paper 3D/1.

10. R.N.Allan, A.Adraktas, J.F.Campbell, "Interactive safety assessment of nuclear reactor systems". Reliability Engineering, 3, 1982, pp.393-410.

11. R.N.Allan, I.L.Rondiris, D.M.Fryer, "An efficient computational technique for evaluating the cut/tie sets and common cause failures of complex systems". IEEE Trans, R-30, 1981, pp.101-109.

12. I.A.Watson, "Review of common cause failures". NCSR Report No. R27, 1981.

13. R.N.Allan, R.Billinton, "Effect of common mode, common environment and other common factors in the reliability evaluation of repairable systems". 7-th Advances in Reliability Technology Symposium, April 1982, Bradford, paper 4B/2.

14. P.D.T.O'Connor, "Practical Reliability Engineering". Wiley, 1985.

15. Y.A.A.Jebril, "Monte Carlo Simulation in Power System Reliability Evaluation". MSc Dissertation, UMIST, 1985.

16. J.M.Hammersley, D.C.Handscomb, "Monte Carlo Methods". Wiley, 1964.

5

■

Design for Reliability

P. D. T. O'CONNOR
British Aerospace
Hertfordshire, UK

INTRODUCTION

A first principle of design must be that the design is inherently reliable, in the expected environment and taking account of the expected production processes. This requires that the designer, in addition to understanding and taking account of the usual design features such as stress, weight, appearance, etc , must also fully appreciate the range of environmental conditions throughout the life of the product, and the range of production methods which will be used in its manufacture. With the very short development timescales now typical of many modern products, and the importance of the design being correct the first time, the traditional approach which led to reliable design through experience and gradual refinement as a result of problems found and corrected is no longer adequate for most modern products. Therefore it is necessary to adopt more disciplined approaches to design, to ensure that the high demands of the modern market for reliability are met. This paper gives an overview of methods which have been developed to ensure that designs are inherently reliable.

THE ENVIRONMENT

For any product, there will be a range of environmental conditions which it will have to withstand during its life in service. The main environmental factors which can affect reliability are:

Temperature

Vibration

Shock

Humidity

Dirt and other contamination

People

Electrical and electronic equipment is further subjected to:

Voltage transients, including static electric discharge.

Electromagnetic effects.

Certain other environmental stresses can affect reliability in certain cases. Examples of these are:

Radiation

Lubricant condition

Altitude or vacuum

Pollution

Salt spray

Fungus

etc.

For each environment, the whole range must be considered, as well as the rate of change when appropriate. For example, rate of change of temperature is often more critical to reliability than is maximum or minimum operating temperature, since cyclic temperature changes can initiate fatigue damage.

Environmental stresses do not operate in isolation, and the combined effects can often be more damaging than a single condition. For example, temperature cycling combined with vibration can lead to faster crack propagation in many stressed components.

People are a special category of the environment, from the perspective of the product. People, whether users, maintainers, or handlers, can generate many of the stresses listed above, and can affect reliability in other ways, often difficult to foresee. People drop and bump products, they pack them incorrectly, they operate controls in the wrong sequences, they forget to fasten panels and they do not always maintain equipment properly. The designer must be aware of all of this, and the product must be designed (and tested) to ensure that it is as robust as practicable against these conditions of treatment.

STRESS ANALYSIS

For a design to be reliable in the range of environments it is likely to have to withstand during its expected service life, it is essential that the design team takes account of all the stresses and combinations of stresses that might be applied. The designers must be aware of the stress application range, ideally written into the product specification.

In order to ensure that all stresses and stress combinations have been properly considered, a formal stress analysis should be performed. The review should be performed by a team, including the designers and by specialists who can advise on aspects such as environmental resistance of materials used, test results, etc. The review should be documented, with a record to show the stress conditions reviewed and the results.

PRODUCTION ASPECTS

Production processes are always subject to variability. The design must take account of these by attention to tolerancing and design for ease of production and test. This requires that the design team are thoroughly familiar with the production processses and their likely variations, and that production engineers work closely with the designers.

COMBINED EFFECTS AND ROBUST DESIGN

The variations of environmental conditions and production variability inevitably combine to affect reliability. Product designs therefore must take account of this.

Tolerance analysis

Tolerance analysis is a statistical method for optimising the tolerances of parts which must fit together or operate together, when the output value or specification is to be met as cost-effectively as possible. Tolerance analysis can also be very useful in optimising yield from production, for example of electronic assemblies.

A simple example of tolerance analysis is the case where a shaft must fit into a hole, both having variation about the nominal diameters. Traditional "worst case" tolerancing would specify that the higher limit on the shaft diameter was lower than the lower limit of the hole diameter, so all shafts would fit all holes. However, if the dimensions were distributed normally, most combinations would not have a close fit. Statistical tolerancing enables us to calculate the probability of a shaft not fitting a hole. The formula for this is:

$$P = 1 - \Phi\left[\frac{D_1 - D_2}{(\sigma_1^2 + \sigma_2^2)^{1/2}}\right]$$

The difference between worst case tolerancing and statistical tolerancing is illustrated in Figure 1. This shows that a slightly larger proportion of combinations will not fit with statistical tolerancing, but that the average clearance will be much less.

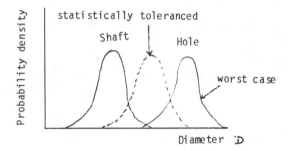

Figure 1. Statistical Tolerancing

This is a simple example to illustrate the point. In practice other factors will have to be taken into account, such as surface roughness and the statistical process control methods used in production.

In many cases, particularly in electronic circuits, the component values and tolerances are set by the component manufacturer. Therefore, the circuit designer cannot apply the statistical tolerancing techniques to the same level of refinement for most component combinations. In such applications it is necessary to assess which are the critical parameter combinations from the tolerance and parameter variation point of view, and to apply statistical tolerancing where such combinations can affect yield or performance. Modern electronic circuit analysis software, such as SPICE, enable this type of assessment to be performed economically and effectively.

Where multiple tolerance possibilities exist, the statistical methods can be extended to cover these. The methods described in chapter   (ANOVA, response surface analysis) can be used to optimise such designs, by showing which are the most significant sources of variation and by enabling these to be controlled at their optimum values.

Robust Design

Robust design is a term used to define a design approach pioneered in Japan by Genichi Taguchi. The "Taguchi Method" combines elements of control theory and statistical design to optimise product designs in relation to their ability to be "robust" to variations in the environment and to production variations. A product can be considered as shown in Figure 2, with the output value (specification) affected by control inputs, and also by external factors such as environmental changes, handling, operation, etc., and by internal factors such as production-induced variations such as component, process and material variations. The designer's challenge is to design the product so that its output value, within the specification limits, is maintained despite the expected range of variation of all the internal and external variations which can affect it.

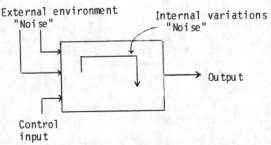

External environment
"Noise"

Internal variations
"Noise"

Output

Control
input

Figure 2.   Design as a Control System

A simple example to illustrate the Taguchi approach is shown in Figure 3. The curve represents the operating characteristic of a component used in a system, with the control value c on the horizontal scale and the output value V on the vertical scale. The specified output value is Vs. In order to achieve this output the control input must be set at cs. If the tolerance on the output is as shown, the tolerance on c must be closely controlled within the limits shown. However, if instead an output value of Vo is used, the control value can be set at co, with much wider tolerance limits sufficing to maintain the output within the same tolerance band as before. A second component with a linear response function is then used to bring the output value back to the specified level Vs.

A design incorporating the approach illustrated above is likely to be:

1. Easier to produce, since the control value will not have to be held to

84

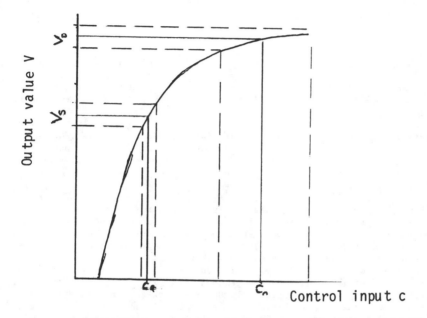

Figure 3. Taguchi Design

such close tolerances.

2. Cheaper to produce, since components and production processes can have wider tolerances.

3. More robust against parameter and process variations, since the specified output will be met over a larger range of input tolerances.

4. More robust against environmental changes, parameter drift in time and other factors that can affect the output parameter value in service. In other words, more reliable.

In a design, the parameter values which merit attention to the Taguchi approach must be identified. Generally, any parameter value which requires to be held to close tolerances in order for the functional specification to be met, expensive components or processes, or components or processes for which high production yield might be difficult to maintain, are candidates for the approach.

The example showed a single source of variation. When multiple sources exist, for example a component tolerance and an environmental change or parameter drift with age or wear, it is necessary to evaluate the combined effect. Analysis of variance (ANOVA) and response surface analysis techniques can be used for this purpose.

The Taguchi approach has been used widely in Japan for several years, and is now generating interest in the West. It seems to be at least part of the reason why the Japanese have been able to create intricate designs and to put them into mass production at low cost and with high reliability. Products such as cameras and lenses, electronic instruments, vehicles, and other consumer electronics products have all been introduced to the market at low cost and with very high quality and reliability. The method provides a very systematic, disciplined and practical way of ensuring that a design is as robust and reliable as possible, and at the same time economic to produce. The Taguchi design approach, coupled with the statistical process control methods to keep production variation within the tolerances identified as neccessary by the analysis, are very powerful techniques when used in combination, in a totally integrated approach to design and production.

DESIGN REVIEW METHODS

Failure Modes, Effects and Criticality Analysis (FMECA)

FMECA is a design review method in which each mode of failure of every component or function within a system is assessed for probability of occurrence, effect of failure, and criticality in terms of successful operation, safety, maintenance, etc. The method is described in US MIL Handbook 1629 and in reference 1. The traditional approach to FMECA, as described in, for example, MIL HDBK 1629, involves tabulating this information on standard forms, and the work can be tedious and expensive. However, computerised FMECA methods are now available, which enable the analysis to be performed much more effectively and economically.

FMECA is a very useful technique for evaluating the safety and reliability of a design, to highlight potential problems early in the development cycle so they can be corrected with the minimum of expense and delay to the project. It requires a disciplined approach, and competent, independent engineers working in close collaboration with the designers.

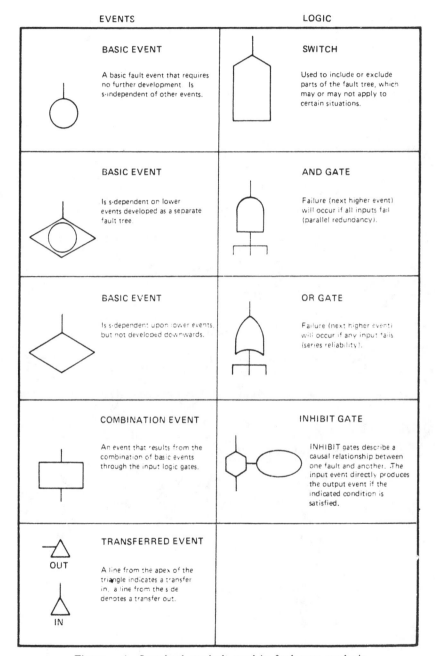

EVENTS

LOGIC

BASIC EVENT

A basic fault event that requires no further development. Is s-independent of other events.

SWITCH

Used to include or exclude parts of the fault tree, which may or may not apply to certain situations.

BASIC EVENT

Is s-dependent on lower events developed as a separate fault tree.

AND GATE

Failure (next higher event) will occur if all inputs fail (parallel redundancy).

BASIC EVENT

Is s-dependent upon lower events, but not developed downwards.

OR GATE

Failure (next higher event) will occur if any input fails (series reliability).

COMBINATION EVENT

An event that results from the combination of basic events through the input logic gates.

INHIBIT GATE

INHIBIT gates describe a causal relationship between one fault and another. The input event directly produces the output event if the indicated condition is satisfied.

OUT

TRANSFERRED EVENT

A line from the apex of the triangle indicates a transfer in; a line from the side denotes a transfer out.

IN

Figure 4 Standard symbols used in fault tree analysis

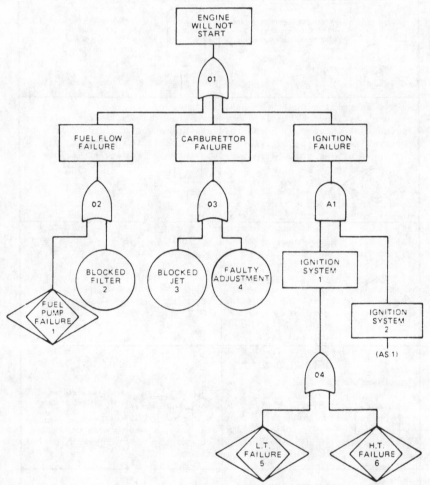

Figure 5 FTA for engine (incomplete) (From ref. 1)

FMECA is also useful for evaluating other design features such as testability using built-in test (BIT) facilities or external testers, and as an aid to preparing diagnostic information.

FMECA can also be performed, related to production processes, to assess the effects of failures of these rather than of components or functions in the product. Such Production FMECAs can be a valuable way of reviewing the production processes in the same disciplined, documented way as the design itself is evaluated, and there can be useful cross-checking between the two approaches.

Fault Tree Analysis (FTA)

FTA is a reliability/safety design analysis technique which starts from consideration of system failure effects, referred to as "top events", and proceeds by determining how these can be caused by single or combined lower level failures or events. It is a top-down analysis, and takes multiple failure modes into account. Standard graphical symbols are used to construct the fault tree picture, by describing events and logical connections. These are shown in figure 4, and a simple FTA is shown in Figure 5. FTA is fully described in reference 1.

FTA is a more powerful method than FMECA for safety analysis, since it can take account of multiple failures, including dormant failures, human failures or incorrect actions, and complex system logic such as redundancy. If probalities are assigned the events, the FTA can be used to evaluate the probability of the defined top event(s) occurring, and the combinations of events which must occur for this to happen. When performed in this way, the evaluation becomes complex for even quite simple systems, and therefore computer programs have been developed for FTA. These evaluate the top event probabilities and identify the single and combination events which would cause them. Some of these programs can also construct the FTA diagrams, using suitable graphics plotters.

FTA is the standard safety analysis method used in industrial contexts such as nuclear power and safety-critical process plant.

Sneak Analysis (SA)

Sneak Analysis (SA) (or Sneak Circuit Analysis, SCA), is a design analysis method which assesses the ways in which a system might operate incorrectly, when no part or function has actually failed. The method was developed for NASA after the Appolo capsule fire, which was caused by a power bus being shorted to earth when a certain system state was reached as a result of switching configurations. In other words, the system failure mode was inherent in the design. The SA method analyses the system topology to identify such "sneak" paths, which might be caused by sequencing, timing, combinations, incorrect warnings or indications, etc. The method is applicable equally to software as to hardware, and can be particularly useful for evaluating the combined hardware-software system.

Because of the complexity of the analysis, computer programs have been developed to enable it to be performed cost-effectively.

FORMAL DESIGN REVIEW

All of the review methods described in the previous section should be properly
documented as part of the design record, for subsequent reference in case of
design changes, problem investigation or as supportive evidence in case of
litigation. They should also be used as inputs to the formal design review.

The formal design review is a meeting at which the design team presents the
design to a panel, chaired by a person who can make decisions, in order to
enable a detailed and independent appraisal to be carried out before the
design is permitted to proceed to the next stage of development or to
production. Design reviews are often based on checklists, appropriate to the
product and the processes involved. The design review panel members include
specialists who can check on particular aspects, such as testability,
environmental resistance, reliability, and ease of production, as well as
design signatories who must approve the design, eg. the quality representative.

It is important that design reviews are set up as key events in the
development of the product, and that they are managed and staffed by competent
people, since they are the opportunity for critical assessment and crucial
decision-making. It is therefore essential that the checklists are carefully
planned, to ask the right questions and to extract constructive responses.
The results of the design analysis techniques described in the previous
section should all be considered at the formal design review.

INTEGRATED DESIGN AND CAD

For modern, complex products to be developed at minimum cost and in good time,
and to be economic to produce, reliable and safe, it is essential that the
design and development process be tackled as a team effort. The team should
include all of the engineering skills necessary, such as industrial design,
component and material specialists, production test engineers, cost engineers,
reliability/safety engineers, etc., in addition to the product design
specialists. The team should be led by an engineer with wide knowledge and
leadership skill. If the full-time attention of certain specialists is not
justified, they should be called in to assist as required, but made to feel
part of the team. This approach to the management of design has been called
integrated engineering.

The advent of computer-aided engineering (CAE) has made the need for an
integrated, team approach to design even more necessary. It also makes it
easier to integrate the separate contributions. For example, there can be a
common design data base, including the design itself, the stress analyses,
reliability and safety analyses, and records of configuration and reviews.
These modern methods and tools, when properly integrated and managed, greatly
increase the capability to create safe, reliable, and competitive products.

REFERENCES

1. P.D.T. O'Connor, Practical Reliability Engineering, J. Wiley & Sons Ltd.,
(2nd. Edition, 1985)

6

■

Reliability of Mechanical Components and Systems

P. D. T. O'CONNOR
British Aerospace
Hertfordshire, UK

INTRODUCTION

Failures of mechanical components and systems occur primarily due to two causes:

1. Shock overload.
2. Deterioration of strength, so that working loads cause failure.

Failures can be due to a combination of these causes, for example high shock loads leading to progressive strength degradation due to fatigue. However, this chapter will treat the two causes separately.

Failures can also occur as a result of factors which are only indirectly or partially related to load and strength interactions, for example damage to hydraulic couplings due to mishandling or incorrect assembly.

The selection of materials and processes are important aspects of design for reliability of mechanical components and systems.

This chapter briefly describes these major facets of mechanical reliability.

SHOCK OVERLOAD

A common cause of failure results from the situation when the applied load exceeds the resisting strength. Load and strength are considered in the widest sense, ie. it might be a mechanical stress such as torque, or an electrical stress such as voltage, or combined stresses such as electrical current and temperature. Strength can be any resisting physical property, such as toughness, hardness, melting point, or adhesion. If the strength exceeds the load, failures should not occur. This is the standard approach to design, in which safety factors are applied related to the expected extreme values of load and strength.

Distributed Load and Strength

For most products, loads and strengths are not fixed, but are statistically distributed, as shown in Figure 1. For normally-distributed variables,

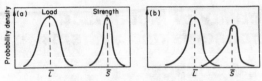

Figure 1 Distributed load and strength: (a) non-overlapping distributions, (b) overlapping distributions

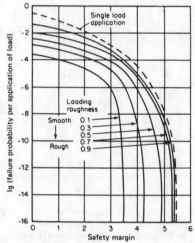

Figure 2. Failure probability–safety margin curves when both load and strength are normally distributed (for large n and $n = 1$) (From ref. 1)

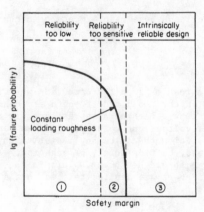

Figure 3. Characteristic regions of a typical failure probability–safety margin curve (From ref. 1)

there is a mean (or median) value $(\bar{L}, \bar{S})$ and a standard deviation (σ_L, σ_S).

When load and strength are distributed in this way, we can define two factors:

1. Safety Margin (SM) = $\dfrac{\bar{S} - \bar{L}}{(\sigma_S^2 + \sigma_L^2)^{1/2}}$

2. Loading Roughness (LR) = $\dfrac{\sigma_L}{(\sigma_S^2 + \sigma_L^2)^{1/2}}$

It can be shown that, when the load and strength are both normally distributed, the reliability of a component for a single load application is given by

$$R = \Phi(SM)$$

where Φ is the standard normal cumulative distribution function.

Example

A component has a strength which is normally distributed, with a mean value of 5,000 N and a standard deviation of 400 N. The load it has to withstand is also normally distributed, with a mean value of 3,500 N and a standard deviation of 400 N. What is the reliability per load application?

The safety margin is

$$\frac{5,000 - 3,500}{\sqrt{(400^2 + 400^2)}}$$

$$= 2.65$$

From tables of the standard cumulative normal distribution function,

$$R = \Phi(2.65) = 0.996$$

For multiple load applications, reliability becomes a function of both safety margin and loading roughness. This complex integral function cannot be reduced to a formula as for the single load case, but can be evaluated using computerised numerical methods. The effects of different values of safety margin and loading roughness on failure probability per load application are shown in Figure 2. This shows the situation for multiple loads, as well as the single load situation.

Since the load applications are statistically independent, reliability over n load applications is given by

$$R = (1-p)^n$$

where p is the probability of failure per load application.

For small values of p,

$$R \approx (1 - np)$$

Thus we can use Figure 2 to derive the value of p, and so evaluate the component's reliability for a given number of load applications. Note that, once the safety margin exceeds a value of 3 to 5, depending on the value of loading roughness, the failure probability becomes infinitesimal. The item can then be said to be <u>intrinsically reliable</u>. There is an intermediate region in which failure probability is very sensitive to changes in distribution parameters, whilst at low safety margins the failure probability is high, as would be expected. Figure 3 shows these characteristic regions.

Note that the above discussion holds for normally distributed load and strength. The same principles apply for other distribution, and similar curves can be derived. See reference 1.

The safety margin and loading roughness allow us, in theory, to analyse the way in which load and strength distributions interfere and so generate a probability of failure. By contrast, traditional deterministic safety factors, based upon mean or maximum and minimum values, do not allow reliability estimates to be made. However, good data on load and strength properties are very often not available, particularly at the extremes of the distributions which are the areas of interest in the reliability context. Other practical difficulties arise in applying the theory, and it is important to appreciate that measured values of load and strength are likely not to be extrapolated with confidence into the future. The large measure of uncertainty shown in Figures 2 and 3 are magnified further by such practical considerations, so the theory must be applied with care and with awareness of its limitations. Nevertheless, it can provide guidance in particular critical design situations. These aspects are discussed in more detail in Reference 1.

STRENGTH DETERIORATION

So far we have taken no account of the possibility of strength reduction with time or with applied loading. The methods described above are relevant only when we can ignore progressive strength reduction, for example if no weakening is expected to occur, or, if weakening will occur, for example due to fatigue, if the item is to be operated only well within the fatigue life.

Causes of Strength Deterioration

The main causes of strength deterioration in components and materials are:

1. Fatigue damage, in which cyclical stresses above a critical value cause cumulative damage and thus weakening.
2. Wear, when surfaces moving in contact are damaged, resulting in higher friction, further damage and failure.
3. Corrosion, when materials (particularly ferrous alloys) are chemically attacked by oxygen ions in water.

There are other causes of strength deterioration which can be important in particular cases. For example:

4. Other chemical attack.
5. Physical change (shape, brittleness, etc.) due to ultraviolet light absorbtion (primarily plastics).
6. Damage due to mishandling.

This chapter will concentrate primarily on the first three causes.

94

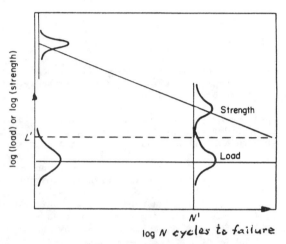

Figure 4. Strength deterioration with cyclic overload (From ref. 1)

Fatigue

Fatigue damage is caused when a mechanical stress is applied to a component, the stress being above a limiting value called the critical stress. The damage is due to internal structural deformations, such as crystal lattice deformations in metals, which do not return to the original pre-stressed condition when the stress is removed. Fatigue damage is cumulative, so that repeated or cyclical stress above the critical stress will eventually result in failure. For example, a spring subjected to cyclic extension beyond the critical stress will ultimately fail in tension. Figure 4 shows this type of situation graphically.

An early empirical model of the relationship between applied cyclic stress and time to failure is Miner's rule:

$$\frac{M_1}{N_1} + \frac{M_2}{N_2} + \ldots \frac{M_k}{N_K} = 1$$

where M_1, M_2,....M_k are the numbers of cycles at a specific stress level, and N_1, N_2,....N_k are the numbers of cycles to failure at those stress levels, as determined by tests, and as shown on the S-N curve (Figure 4).

Other relationships have been developed. However, since times to failure in fatigue are very variable (an order of magnitude or more), the use of a simple relationship such as Miner's rule is usually sufficient for design assessment.

Analyses of times to failure in real or simulated test situations can be performed using Weibull probability plots. Weibull distributions of times to failure show a positive failure-free life (γ) and slope (β) values greater than 1 (typically 2-3.5), ie. an increasing hazard rate with failures starting after the safe life interval. This life is sometimes quoted as the "B-life", ie. the life interval at which B% of the items have been shown to have failed on test. B-lives are also used to define the lives of components subject to wear, eg. bearings.

Design Against Fatigue

Design to protect against failure due to fatigue involves:

1. Knowledge of material fatigue properties.
2. Control of stress distributions within the material, for example by obviating stress raisers such as sharp internal corners.
3. Design for safe life under the expected operating conditions.
4. Design for "fail safe", ie. the load can be taken by another component or the effect of failure is otherwise mitigated, until the failed component can be detected and repaired or replaced. This approach is common in aircraft structural design.
5. Design for ease of inspection to detect fatigue damage (cracks).

Maintenance of Fatigue-prone Components

It is very important that critical components subject to fatigue loading can be inspected to check for crack initiation and growth. Maintenance techniques for such components include:

1. Visual inspection.
2. Non-destructive test (NDT) methods, such as dye penetrants, X-ray, and acoustic emission tests.

3. Scheduled replacement before the end of the fatigue life.

The scheduling and planning of these maintenance techniques must be based upon knowledge of the material properties (fatigue life, crack propagation rates, variability), the load duty cycle, the effect of failure, and test data.

WEAR

Wear Mechanisms

Wear is the removal of material from the surfaces of components as a result of their movement relative to other components or materials. Wear can occur by a variety of mechanisms, and more than one mechanism may operate in any particular situation. The main wear mechanisms are described below.

Adhesive Wear occurs when smooth surfaces rub against each other. The contact load causes interactions between the high spots on the surfaces and the relative motion creates local heating and dragging between the surfaces. This results in particles being broken or scraped off the surfaces, and loose particles of wear debris are generated.

Fretting is similar to adhesive wear, but it occurs between surfaces subject to small oscillatory movements. The small movements prevent the wear debris from escaping from the wear region, so the particles are broken up to smaller sizes and might become oxidised. The repeated movements over the same parts of the surface also result in some surface fatigue, and corrosion also contributes to the mechanism.

Abrasive wear occurs when a relatively soft surface is scored by a relatively hard surface. The wear mechanism is basically a cutting action often with displacement of the soft material at the sides of grooves scored in the soft material.

Fluid erosion is caused to surfaces in contact with fluids, if the fluid impacts against the surfaces with sufficient energy. For example, high velocity fluid jets can cause this type of damage. If the fluid contains solid particles the wear is accelerated. Cavitation is the formation and violent collapse of vacuum bubbles in flowing liquids subject to rapid pressure changes. The violent collapse of the vacuum bubbles on to the material surfaces causes fluid erosion. Pumps, propellors and hydraulic components can suffer this type of damage.

Corrosive wear involves the removal of material from a surface by electrolytic action. It is important as a wear mechanism because other wear processes might remove protective films from surfaces and leave them in a chemically active condition. Corrosion can therefore be a powerful additive mechanism to other wear mechanisms.

Methods of Wear Reduction

The main methods of wear reduction are:

1. Minimise the potential for wear in a design by avoiding as far as practicable conditions leading to wear, such as contact of vibrating surfaces.
2. Selection of materials and surface treatments that are wear-resistant or self-lubricating.

3. Lubrication, and design of efficient lubricating systems and ease of access for lubrication when necessary.

When wear problems arise in use, an essential starting point for investigation is examination of the worn surfaces to determine which of the various wear mechanisms, or combinations of mechanisms, is involved. For example, if a plain bearing shows signs of adhesive wear at one end, the oil film thickness and likely shaft deflection or misalignmet should be checked. If the problem is abrasive wear the lubricant and surfaces should be checked for contamination or wear debris.

In serious cases design changes or operational limitations might be needed to overcome wear problems. In others a change of material, surface treatment or change of lubricant might be sufficient. It is also important to ensure that lubricant filtration, when appropriate, is effective.

Maintenance of Systems Subject to Wear

The life and reliability of components and systems subject to wear are very dependent upon good maintenance. Maintenance plans should be prepared, taking into account cleaning and lubrication requirements, atmospheric and contamination conditions, lubricant life and filtration, material properties and wear rates, and the effects of failure.

CORROSION

Corrosion affects ferrous and some other non-ferrous engineering metals, such as aluminium and magnesium. It is a particularly severe reliability problem with ferrous products, especially in damp environments. Corrosion can be accelerated by chemical contamination, for example by salt in coastal or marine environments.

The primary corrosion mechanism is oxidation. Some metals, particularly aluminium, have oxides which form as very hard surface layers, thus providing protection for the underlying material. However, ferrous alloys do not have this property, so oxidation damage is cumulative.

Galvanic corrosion can also be a problem in some applications. This occurs when electromotive potentials are built up as a result of dissimilar metals being in contact and conditions exist for an electric current to flow. This can lead to the formation of intermetallic compounds and the acceleration of other chemical action. Also, electrolytic corrosion can occur, with similar results, in electrical and electronic systems when induced currents flow across dissimilar metal boundaries. This can occur, for example, when earthing or electrical bonding is inadequate.

Stress corrosion is a combination of cumulative stress and corrosion damage. It is caused by corrosion precipitating surface weaknesses, leading to crack formation. Further corrosion and weakening occurs at crack tips, where the metal is in a chemically active state and where the high temperatures generated accelerate further chemical action. Thus the combined effect can be much faster than either occurring alone.

Design methods to prevent or reduce corrosion include:

1. Selection of materials appropriate to the application and the expected

environments.

2. Surface protection, such as anodising for non-ferrous metals, painting, metal plating (galvanising, chrome plating, etc.), and lubrication.

3. Other environmental protection, such as the use of dryers or desiccators.

4. Avoidance of situations in which galvanic or electrolytic corrosion can occur.

5. Awareness and avoidance of conditions likely to generate stress corrosion.

Correct maintenance is essential to ensure the reliability of corrosion-prone components. Maintenance in these situations involves ensuring the integrity of the protective measures described above. Since corrosion damage is usually extremely variable, scheduled maintenance should be based upon experience and criticality.

OTHER MECHANICAL FAILURE MECHANISMS

There are many other mechanisms of failure of mechanical components and systems. Some of these can be exacerbated by the mechanisms described above. Examples of these other mechanisms are:

1. Backlash in controls, linkages and gears.
2. Incorrect adjustments on valves, metering devices, etc.
3. Seizing of moving parts in contact, such as bearings, due to contamination, corrosion, or physical damage.
4. Leaking of seals.
5. Loose fasteners, due to incorrect tightening, wear, or inadequate locking.

The designer must be aware of these and other potential causes of failure, and must design to prevent or minimise their occurrence. Appreciation of "Murphy's Law" (if a thing can go wrong, it will) is essential, particularly in relation to systems which are maintained and which include other than simple operator involvement.

MATERIALS

Selection of appropriate materials is an important aspect of design for reliability, and it is essential that designers are aware of the relevant properties in the application environments. With the very large and increasing range of materials available to designers this knowledge is not easy to retain, and designers should obtain material data and application advice from suppliers as well as from handbooks and other data bases. A few examples of points to consider in selecting engineering materials for reliability are given below.

Metal Alloys

1. Corrosion environment, compatibility.
2. Surface protection methods.
3. Electrochemical (electrolytic, galvanic) corrosion if disimilar metals in contact.

Plastics, Rubbers

1. Resistance to chemical attack from materials in contact or in atmosphere

(lubricants, pollutants, etc.)
2. Temperature stability (dimensional, physical).
3. Sensitivity to ultraviolet radiation (sunlight).

Ceramics

Brittleness, fracture toughness.

Composites

1. Impact strength.
2. Erosion.
3. Directional strength.

For all materials, it is of course necessary to take account of all of the properties relevant to the application and the likely environments. These properties include strength, fatigue properties, wear and corrosion resistance, etc.

PROCESSES

Designers must be aware of the reliability aspects of the manufacturing processes which might be used for the product. For example, machining processes always involve working to tolerances, and material properties such as fatigue life can be affected by machining. Heat treatment also affects the properties of many materials, particularly metals and ceramics. Surface treatment, such as metal plating, shot peening, chemical treatment and painting, can greatly affect reliability and endurance in particular cases.

REFERENCES
1. O'Connor, P.D.T., Practical Reliability Engineering (2nd. edition), Wiley, 1985.

2. Carter, A.D.S., Mechanical Reliability (2nd edition), Macmillan, 1986.

3. Besuner, P.M., Harris, D.O. and Thomas, J.M., (editors), A Review of Fracture Mechanics Life Technology, NASA Report 3957, 1956.

4. Hertzberg, R.W., Deformation and Fracture Mechanics of Engineering Materials, J. Wiley, 1976.

5. Collins, J.A., Failure of Materials in Mechanical Design, J. Wiley, 1981.

6. Sayles, R.S., Webster M.M., and MacPherson P.B., The Influence of some Tribological Problems in Mechanical Reliability, Proc. Seminar on Mechanical Reliability in the Process Industries, IMechE, London, 1984.

7. Lipson, C., Wear Considerations in Design, Prentice-Hall, 1967.

8. Uhlig, H.H., and Revie, R.W., Corrosion and Corrosion Control (3nd edition), J. Wiley, 1985

9. Brostow, W., and Corneliussen, R.D., Failure of Plastics, Hanser, 1986.

10. Crane, F.A.A, and Charles, J.A., Selection and Use of Engineering Materials, Butterworths, 1984.

7

Reliability of Electronic Systems

P. D. T. O'CONNOR
British Aerospace
Hertfordshire, UK

INTRODUCTION

This chapter briefly surveys the methods available to ensure the reliability
of electronic systems. The main characteristics of electronic systems which
distinguish them from non-electronic systems and products is that wearout
failure mechanisms are very uncommon. For the great majority of modern
electronic devices and manufacturing processes (solder, wire bonding, etc.)
the only destructive failure modes of significance are those due to overstress
or the effects of defects in production of devices and systems. These lead
predominantly to constant failure rates (overstress failures) or decreasing
failure rates (quality defects) in repairable systems. The preventive
measures against these predominant failure modes are primarily protection
against overstress and quality control measures in device and system
production to control variation.

There are other failure mechanisms which affect electronic systems. Common
among these are electromagnetic interference (EMI), primarily affecting
digital systems, drift of component parameter values over time, and various
mechanical failures such as connector breakage, solder joint failures,
corrosion, etc.

However, modern electronic systems can be designed and produced to be very
reliable indeed, over very long periods and in severe environments. The
components and processes are very robust, and careful design, component
selection and quality control can result in intrinsically reliable systems.
On the other hand, lack of attention to any of these aspects can lead to very
poor reliability.

DESIGN

There are three main elements of electronic system design which affect
reliability. These are:

1. System design, taking account of input and output specifications,
environment, constraints of size, power, etc., and other performance-related
aspects.

2. Selection of the right components, taking account of component parameters,
tolerances and environmental capability.

3. Process design, to ensure that the production processes to be used will not adversely affect reliability and will be optimised for cost effectiveness.

These aspects are not of course unique to the design of electronic systems, but certain aspects deserve special emphasis for application in this area. Each will be discussed in more detail below.

General Aspects

The design must obviously meet the performance objectives as specified. These will normally include input and output parameter values and logic, requirements for speed, accuracy, robustness, etc. Good design practice, based on sound training and experience, is the first requirement to ensure that designs are correct, without having to go through expensive and time-wasting revisions. Therefore training, supervision and review are prerequisites for reliable design, and it is very difficult and expensive to rely on subsequent analysis and test to discover design errors. As with all reliability and quality control, prevention is much better than cure, so it is difficult to over-emphasise the need for attention to these aspects, including the application of sufficient time and other resources.

The methods described in Chapter 5, covering design analysis and review techniques, apply equally to the design of electronic systems.

Protection

It is necessary to protect circuits and components against externally and internally generated stresses and interference which might cause failure. The main stresses and the relevant protection methods are described below.

Transient electrical stress. High transient electrical stresses, which can be generated by switching of inductive loads, power supply transients due to switching or lightning, and electrostatic discharge from operators, maintainers and people handling the product during production, packaging, shipment and storage. Modern electronic components are prone to damage from such transients, since they are often very small and therefore have little heat sinking capacity. Also, since most modern systems, particularly digital systems, operate at low voltage levels (typically 5V), susceptibility to damage from transients which can be of the order of kilovolts in the case of electrostatic discharge, is high.

Methods to provide protection against transient voltages include:

1. Capacitors to absorb high frequency/fast rise time transients. This method is commonly used in digital systems, in which all component inputs and outputs which connect to PCB edge connectors are protected in this way by, typically, 0.01μF solid tantalum "buffer" capacitors.

2. Resistors placed between inputs/outputs and external connections, to reduce the value of transient voltages and to prevent excessive load being applied in the case of a short of the output to ground.

3. Isolation, using, for example, opto-couplers, to physically separate sensitive components from damaging transients.

4. Earthing and bonding, to ensure that "earth loop" currents are not generated.

5. Derating (see below).

Thermal design. Electronic components generate heat in operation, and the combined heat generated by many components in a system, added to any other heat input such as external ambient temperature and solar radiation can result in component temperatures approaching or exceeding maximum rated values. Typically, these are in the range 80C to 150C. Whilst thermal design to keep temperatures below these levels is not usually a problem in fairly simple domestic and commercial electronic systems, the complexity, high packing density and other constraints make thermal design of more complex systems an important aspect of overall design for reliability. Typical situations in which thermal design becomes important are in avionics, computers, complex test equipment, and most military equipment.

The main reasons for setting upper temperature limits on component operation are that parameter values often vary with temperature, and at temperatures above the maximum rated the suppliers no longer guarantee that parameters will be within specification. Also, age-related parametric drift is accelerated by high temperature, so circuit elements can drift out of specification within the working life of the system if not adequately cooled.

Thermal protection methods include:

1. Heat sinking for components which dissipate considerable power. The components are mounted on heat sinks which absorb and radiate heat.

2. The use of a thermal conduction plane within or on printed circuit boards. This may also serve as a common earth. The conduction plane conducts heat back to the edge connection mechanical interface, and thus away from the heat generating components to the outside of the unit via cooling units or the enclosure walls.

3. Fans, used either to circulate air to improve heat flow to the enclosure walls, or to direct cooling air through the enclosure.

4. For severe heat dissipation requirements, for example for high power devices and stacked high-speed digital devices in large computers, liquid medium rather than forced air cooling is employed.

Electromagnetic interference (EMI). EMI is a major design reliability problem in digital systems. It is also important in the design of other electrical and electronic systems, since they can emit electromagnetic radiation which can cause interference in nearby systems which are susceptible. Therefore EMI must be tackled both by preventing emission and by protection of susceptible systems.

EMI can cause digital systems to malfunction due to conductors acting as aerials and picking up the external electromagnetic signals, superimposed on the digital data. This can corrupt the digital data. In other systems EMI can cause intermittent or inadvertent functioning of devices such as sensors or relays.

Sources of EMI include:

1. Electric motors (commutators, brushes).

2. Spurious or harmonic emissions from amplifiers, tuners, etc.

3. Radiation from high-energy devices such as spark plugs, transmission lines, transformers, radars, other transmitters, etc.

4. Discharge of electrostatic energy.

5. High speed digital data in conductors lying close to and parallel to other signal conductors.

Note that EMI can be caused by emissions external to and within the system being designed. As digital systems operate at increasingly high speeds (currently up to 50MHz), emission and susceptibility are both increased.

Electromagnetic emissions from equipment are controlled by several national standards, which state limits of power and frequency which may be emitted from systems while operating.

Protection against EMI includes the following:

Emission:

1. Supression.

2. Screening.

Susceptibility:

3. Screening.

4. Filters, to attenuate specific unwanted frequencies.

5. Buffering, isolation and earthing, as described above.

6. Careful design to avoid mutual interference by adjacent components and conductors. This includes considerations of layout, packaging, internal screening and bonding, etc.

Derating. Derating is the term applied to the application of components in such a way that the maximum stresses which will be imposed in operation will never exceed the their specified maximum stress levels. The important stresses for electronic components are temperature and electrical stress (power, voltage or current, as appropriate).

The objectives of stress derating are:

1. To ensure that substandard components which nevertheless pass operating performance tests will not be highly stressed in use, and will therefore be less likely to fail.

2. To reduce parameter drift.

3. To protect against transient overstress.

4. To protect against errors or incorrect assumptions in design or environmental conditions.

The derating to be applied to particular components depends upon the application. Derating guidelines are given in References 1 and 2, and in most component manufacturers' data books.

Since electronic components are available mainly as standard devices, most applications will result in stress values being derated without additional design effort or cost. However, it is important to check that stressed components are adequately derated, and a formal design review check is often imposed. Electronic CAD systems now include appropriate stress analysis routines.

Mechanical Protection. Obviously electronic systems must be designed to withstand the mechanical and atmospheric environments in which they might operate or be stored and transported. Therefore the designers must take account of the effects of mechanical shock, vibration, humidity, and the other environmental stresses that can affect the system. Chapter 6 covers these aspects, but there are some features of most electronic systems which deserve particular mention.

Heavy components (transformers, heat sinks, etc.) should be supported mechanically, rather than relying entirely on the solder connections for mechanical support. Solder has poor fatigue properties, and vibration, shock or thermal cycling can cause fracture leading to open/intermittent circuits or even the component becoming separated. The positions in which heavy components are located can also be critical in relation to vibration resonance and nodes, leading to rapid mechanical failure or circuit intermittencies.

Connectors and cables must be carefully positioned and supported. Connector failure and intermittency is one of the major causes of electronic system failure, and insufficient attention to this mundane aspect of design is often the reason. Connectors and cables are subject to user and maintainer disturbance, mechanical shock and vibration, and connectors are adversely affected by moisture and dirt. Therefore special attention should be given to positioning, routing, and support.

Packaging and protection methods must be selected and designed to provide the appropriate degree of environmental protection.

COMPONENT SELECTION

Introduction

Selection of the right components for the application is a very important aspect of electronics design for reliability. In most circuit applications the required functions can be performed by a range of component types, from different suppliers. Furthermore, the quality of the components purchased can be very variable. It is therefore necessary to evaluate the options to select the components most suitable for the application, taking account of reliability-related factors such as stability, failure modes, environmental robustness, precision, and performance parameters which might affect system function in adverse conditions.

Selection of Integrated Circuits

Selecting the right integrated circuits for a new design requires specialist knowledge of this rapidly evolving technology. The basic options, all affecting reliability, are described below.

Technology. The main technology choices for most applications are bipolar (transistor-transistor logic- TTL) or metal oxide semiconductor (MOS). The main differences are that bipolar devices are faster in operation but dissipate more heat, whereas MOS devices (usually complementary MOS, or CMOS, in which both p- and n-channel field effect transistors are diffused on the same silicon wafer) are slower (though the speed gap has been closing), and dissipate very little heat, since there is negligible current consumption by gates once they have switched. Consequently CMOS-based systems require very low power, and the technology predominates in applications in which power supply is limited, such as portable equipment (calculators, watches, etc.). Of course the low heat dissipation of CMOS contributes to reliability, by making thermal design easier or not a problem.

CMOS devices are less sensitive to EMI than are most bipolar devices, since they can operate over larger ranges of supply voltage. This "noise immunity" makes their application popular in industrial control systems, where there might be high background EMI levels, and in which control might be critical in terms of safety or cost. CMOS-based systems also do not require regulated power supplies.

"Latch-up" is a destructive failure mode of CMOS devices, which results from the action of parasitic transistors within the junction structure. These can act as amplifiers under conditions of high transient voltage, for example power supply transients, circuit oscillation, or EMI/ESD. It is the susceptibility of CMOS to latch-up which causes the technology's relative sensitivity to damage from electrostatic discharge or other transient voltage peaks. Latch-up can also be caused by applying signals to the circuit before it is powered. Latch-up usually causes permanent damage, with the device gate latched to the supply voltage until the device is destroyed. Latch-up is less severe a problem with modern devices, due to the incorporation of protective structures in the diffused silicon.

Standard vs. application-specific. Until fairly recently nearly all design involving the use of microelectronic devices relied on standard components, with the designer's task being that of planning the interconnections between them and other components. Standard components are still widely used, particularly for functions such as microprocessors, memory, multiplexers, etc. However, there is an increasing trend towards the use of application-specific ICs (ASICs), in which the required functions appropriate to a particular application are implemented on customised devices. ASICs can be fully customised, ie. the whole IC design is specific to an application, or part-custom (gate array, cell array or uncommitted logic array- ULA), in which standard gate or cell patterns are customised for an application by design of customised patterns and interconnections. The decision on which option is best for a particular design must be based primarily on economic grounds, since ASICs are more expensive to design and buy in small quantities than are standard devices to make up the same functions. However, more and more products now rely on ASICs for much of their functionality, since they can offer considerable savings in space and power consumption, as well as cost over large enough quantities.

The design of ASICs requires the use of computer-aided design (CAD) and the necessary expertise. They can be designed in-house, and the design information passed to the ASIC manufacturer, or the ASIC manufacturer can work from specifications and perform the design work.

The use of ASICs requires that reliability aspects previously under the control only of the device manufacturer now become controllable by the system designer. He must take account of his requirements for reliability features such as ESD susceptibility, redundancy, power dissipation, testability (see later), etc., whether the design is created in-house or by the ASIC manufacturer. Some of these aspects might be covered in the design rules built in to the CAD system, so the designer must check these carefully, as they might differ between different suppliers.

Packaging. ICs have for long been supplied primarily in the dual in-line package configuration, in either hermetic (metal or ceramic) or plastic packages. Recently, with increasing demand for higher packaging density for modern products, ICs have become available in smaller packages, with smaller spacing between the external connectors. The main development has been in surface mount devices (SMDs), which are small packages with the leadouts on a 0.005 inch or 0.0025 inch spacing around the periphery of the package (leadless chip carrier- LCC), or with the leadouts taken to an array of pins on the underside of the package (pin grid array- PGA). Other packaging options also exist, and even higher density methods such as tape automated bonding and chip-on-board are under intense development.

The reliability implications of packaging technology are important. The DIP makes connection to the circuit via leads which are soldered into holes in the printed circuit board. These provide good mechanical as well as electrical properties. However, the SMDs rely on solder connection to the surface of a substrate (ceramic or epoxy PCB), and the mechanical support provided is not as reliable, particularly for devices and equipment which operate over wide temperature ranges or many temperature cycles, since this can result in fatigue failure of the joints in shear. Also, control of the solder process becomes more difficult, since the connections are much smaller, and, in the case of PGAs, not visible for inspection. SMD technology has been developed for automated component placement and soldering, and manual methods useable with DIPs are not practicable.

Another aspect of IC packaging with reliability implications is whether to use plastic encapsulation or hermetic sealing for the device and its connections. Hermetic packaging, consisting of ceramic or metal containers with the IC and its lead frame sealed into a cavity in which the atmosphere is usually dry nitrogen, offer a high degree of protection against humidity and atmospheric contamination. However, because of the lower costs, plastic moulding, using epoxy or silicone encapsulants, is a lower-cost option for most high-volume standard ICs. Unfortunately the plastic encapsulating materials are not impervious to moisture, and water can also reach the chip surface and bond connections along the plastic-metal interfaces of the leadouts. Moisture penetration can cause failure due to corrosion of the aluminium metalisation or of the bonds. Because of this, plastic encapsulation has not been favoured for high-integrity systems, or for products which must operate or be stored for long periods or in hostile environments. For example, their use has generally been prohibited in military equipment.

However, there have been considerable improvements in the methods used in plastic encapsulation of ICs. Control of the constituents and purity of the compounds has resulted in much lower levels of ionic contaminants, so that the effect of moisture ingress is reduced. Improvements have also been made in the passivation process, leading to fewer cracks and pinholes through which moisture can attack the active surface of the chip. These and other improvements have greatly reduced the risk of using plastic encapsulated ICs, though it is important that close controls be maintained on specifications, testing and processes when such devices are used in severe or high-risk applications. Chapters 8 and 9 cover these aspects in detail.

General Selection Guidelines

To obtain reliable performance from any electronic component, it is essential to take account of the complete range of application and environmental conditions that will apply. Often designs prove unreliable because of inattention to such details, or problems are discovered during development testing which require expensive and time-consuming redesign.

Among the factors to consider are:

1. Parameters which might cause unwanted circuit behaviour. For example, stray capacitance might cause resonance, or insufficient temperature stability might cause circuits to drift out of specification.

2. Suitability for the application. For example, choosing a relay for an application in which the normal operating mode is with contacts' closed and passing a low current would involve some different considerations than choosing one for a high frequency, short duty cycle.

3. Environmental conditions. The application environment can seriously affect component performance and life. The use of plastic encapsulated ICs has been discussed above. Other environmental factors, such as electromagnetic radiation, humidity, contamination and vibration must all be considered. Components often at risk from the environment are connectors, switches and keyboards. Many electronic systems suffer unreliability due to insufficient gold plating on connectors and the use of cheap switches and keyboards.

4. Quality. Components must be selected from sources which can assure that the specified parameters will be achieved consistently. Quality assurance in production will be covered later.

In appropriate circumstances reliability and quality assurance can be obtained by selecting components which comply with recognised national or international standards. These standards provide consistent specifications, quality assurance provisions, and test provisions, and they are supervised by national quality authorities. For example, in the UK British Standard (BS) 9000 components can be used, and the international specification system is operated by the International Electrotechnical Commission. In the USA military specifications can be used. These systems must be used in some situations, for example in military equipment contracts. Components specified according to these systems are sometimes more expensive, but the higher component costs can be more than offset by the savings of higher quality in production and reliability in service.

Since the technology of each component type is complex and dynamic, it is advisable to discuss every application with the potential suppliers' applications engineers, who should be able to advise on possible problems and solutions, circuit design considerations, etc. Total reliance on specification data and written application notes might lead to particular problems being overlooked. Also, since much modern circuit design is performed on CAD, the limitations of electronic CAD software in modelling all parameters of a device must be allowed for.

Process design is the design of the production processes that will be used during the manufacturing phase. Process design is important not only to optimise production costs and to ensure that there is a problem-free transition to production, but it can also greatly affect reliability. For electronic designs the main process design considerations are the assembly and test operations.

Modern electronic production can involve a considerable degree of automation, and this greatly affects quality and reliability, as well as productivity. The main areas for automation have been, first, machine soldering in place of hand soldering, so that complete PCB assemblies could be soldered by being passed through a solder wave. Automatic component insertion was the second major change, resulting in fewer damaged or misplaced components. Of course hand operations for assembly and soldering are still used, for example for low-volume products and for sub-systems such as backplanes and some final production operations, but the increasing flexibility of automatic systems leads to ever wider application. Recently the advent of surface mounted components has necessitated the introduction of a new range of automation, since these components cannot be mounted and soldered by hand at the rates and to the levels of accuracy and quality required. Robotic placement machines and in-line infra-red or vapour-phase soldering systems have been developed for the SMD era, and their use is spreading rapidly.

Automated production facilities can quickly produce large quantities of defective products if any of the processes are not correctly set up and monitored. Because of the very precise control required for component placement, solderability and solder system performance, it is essential that the production processes are thoroughly evaluated during the process design and development phase, before production commences. For mature production lines, for which design rules based on experience have been developed, each design does not need the same degree of process evaluation, but care must be taken whenever different component package types or other changes are being considered. Since the electronics packaging and mounting technologies are in a state of rapid development at present, production process evaluation is essential for most modern designs.

The circuit layout, ie. PCB track layout and component placement, should take account of the soldering methods likely to be used, since design features can affect the performance of the solder system. For hand solder, there must be good access and visibility for all solder points. For wave soldering, the disposition of heat-sinks and PCB track orientation can affect the performance of the soldering system.

Design for test is a key element of process design for modern electronics. Due to the complexity of modern digital ICs, whether standard or ASIC, and the number of such devices in many systems, the cost of testing can be very high, involving large capital outlay on automatic test equipment (ATE). Since the performance of even the best ATE in terms of test speed and test coverage (ie. the proportion of potential failures which the ATE can detect and isolate) depends upon the circuit design, it is important for both production cost and reliability reasons that digital systems (ASICs and assembled systems) are designed to be tested as quickly and comprehensively as practicable. The designer should be aware of the types of ATE which will be used in production. He should also be fully conversant with relevant testability design practice, and be motivated to apply them. It is easy for testability features to be neglected until after designs have been passed to production, since basic performance can be demonstrated on designs which are not easily testable, so a disciplined approach, with reviews, is necessary.

Methods for ensuring that the design is "robust" in relation to production variation (component parameter variation, process variation) as described in Chapter 5, should be applied to all aspects of the system design which require control and optimisation. These methods can significantly influence production costs and reliability.

DEVELOPMENT TESTING OF ELECTRONIC SYSTEMS

No system development is complete without a carefully planned test programme to ensure that the system meets its performance and reliability requirements. Development testing for reliability is covered in Chapter 14.

PRODUCTION

Control of the production processes is, after the basic technology challenges, the most important factor in determining the reliability of modern electronic products. The very rapid advances in electronic components and systems, achieved with steadily decreasing costs, increasing reliability and hence enormous acceptability and market growth, would not have been possible without revolutionary improvements in quality control methods and equipment. For example, the scope for a single or very few defects on a modern very large scale integrated circuit or a system comprising many such devices with the necessary interconnections and peripherals is so large that methods had to be developed to ensure that the manufacturing processes could be operated to levels of quality approaching continuous perfection. The general methods and philosophy of the modern approach to quality control are described in Chapter 17. In this section the methods applicable specifically to electronics will be described briefly.

Electronic Components

It soon became apparent that previously acceptable levels of quality, typical of components produced up to the period shortly after the second world war, in which quality levels could be monitored by statistical sampling techniques such as US Military Standard 105 (British Standard 6001), were unacceptably low for electronic production, in which large numbers of components were being used. These methods are capable of ensuring that proportions of defectives do not exceed figures of the order of one percent, to predetermined statistical confidence levels. In a system of, say, 1000 components, even defect proportions of the order of 0.1 percent are obviously too high. The "lot tolerance percent defective" (LTPD) method, described in, for example, US Military Standard 38510, was introduced in an attempt to tighten up on the quality of delivered components, by reducing the statistical risks as seen by the consumer. However, this approach did not remove the fundamental problem of assuring that proportions defective were really of the order of 0.001 percent or lower. At these levels statistical sampling becomes irrelevant, and attention must be focussed on continuous improvement methods, with no measured level of quality less than perfection being considered "acceptable" or "tolerable". One scheme which has become popular is the "parts per million" (ppm) approach, in which the customer records the proportion defective against a ppm level agreed with the supplier, who agrees to take action necessary to meet the agreed figure. No statistics beyond a measured average are involved, and the action taken in the event of failure to meet the target is co-operative investigation and process improvement with both the user and the supplier being involved. For example, it is quite common for a perceived problem with a component to be due to a test or assembly method being used by the equipment manufacturer, and close co-operation might be necessary for the source of the problem to be uncovered. A key element of ppm schemes is co-operative effort towards continual improvement, rather than action being taken only when statistical "fail" criteria exist. Typical ppm levels currently achieved are 500 (0.002 percent) for complex new VLSI devices to less than 50 for simple standard components. Increasingly component manufacturers are offering "zero defects".

In order to exploit the very high quality and reliability of modern electronic components is is necessary for users, that is, system designers, purchasing staff and production quality control staff to have a good understanding of the test and other quality assurance methods applicable. The failure mechanisms described earlier and in Chapters 8 and 9 and   are inherent in the production and handling processes if there is any relaxation or mistake, and it is easy for the unwary or unprepared buyer to take delivery of a batch of components which is of substantially lower quality than expected.

A further important factor to be considered is component testing. The question whether or not to test electronic components before placement is an important one involving production economics and reliability. Generally, complex components can be thoroughly tested only when not in the circuit. However, component testers for VLSI components are very expensive, and furthermore it is arguable that exhaustive component test is not essential so long as it is possible to test the assembled product thoroughly. A further factor is the expected quality level of the components: if the proportion defective is very low, it is more economic not to test them but to repair defective assemblies. With the dramatic improvements in quality of modern complex components the balance has swung in favour of the no-test decision for components. On the other hand, the capability to test components can be very useful for evaluating new components, particularly ASICs, and for failure analysis, as well as for batch testing during production when there is uncertainty about the quality of a particular component.

"Burn-in" is a test method which has been developed for electronic components and systems, with the objective of causing weak components or assemblies to fail under accelerated stress conditions, which however do not damage good items. Burn-in is based on the principle that in a population of components or assemblies there might be defects which will lead to subsequent failure in use, but which will not cause the item to fail functional tests in the factory. Such defects in electronic components and assemblies cannot usually be detected by other means, such as visual inspection. Therefore burn-in can be a very effective technique for reliability improvement, and it is widely applied to complex components such as integrated circuits, both by component manufacturers and by users, and to electronic assemblies.

Standard burn-in methods as applied to microcircuits are described in documents such as US Military Standard 883 and British Standard 9400. Typically this involves up to 168 hours of operation at 125 C, with applied electrical stress. Variations on these basic methods have been developed, particularly relevant to the more complex modern devices such as large memories and application-specific ICs, involving dynamic stressing and functional test during the burn-in period. These methods are necessary because the long time necessary to perform functional tests of such devices and the high cost of the automatic test equipment used make it worthwhile taking advantage of the relatively long burn-in time to perform functional tests on large numbers of devices simultaneously while under stress. Furthermore, the tests can be tailored to the specific application of the devices, including input and output patterns.

Because of the high capital and operating costs involved, many companies use independent test resources for this work, while others have set up central facilities to serve a number of locations.

Another approach which is becoming increasingly popular for users of large quantities of components is the "just in time" or JIT approach (also called "dock-to-stock"). In this system, pioneered in Japan in industries such as automobile and electronics manufacture, components and sub-assemblies are delivered by the suppliers direct to the production line, exactly when required. There are no factory stocks, receiving inspection or test, and therefore no lead time. The economic benefits are obvious, since working capital is greatly reduced. However, the system relies on very high quality standards being maintained. The supplier is responsible for all inspection and test necessary to ensure that the components can be used as delivered, and a very close engineering relationship is necessary between the system manufacturer and the suppliers. JIT is usually combined with a single-vendor policy, with whom a close partnership is established during product and process design and development.

System Manufacture

Assuring high quality of electronic system production involves primarily:

1. Use of high quality electronic components, as described above.

2. High quality assembly (PCB/substrate manufacture, component placement, soldering, final assembly operations).

3. Effective testing.

4. Assembly burn-in.

These aspects will be considered in turn.

Quality Control of Processes

The basic quality control management and motivational techniques described in Chapter 17, coupled with good production engineering as applicable to the methods and processes employed, will assure high quality production of electronic systems. It is particularly important that a very responsive data collection and analysis system is used, to collect and display defect trends on components and processes. Computerised systems are available, linked directly to automatic test equipment, and these enable trends to be quickly isolated.

Production Testing

Adequate testing is essential to ensure that the manufactured product meets the design performance specification. However, testing of electronic equipment can be expensive, so it is important that the right balance is struck. For example, for complex electronic equipment the cost of test can be up to 30% of the total production cost.

Ideally, if all components and processes are defect-free, there is no need to test. However, even at the high standards achieved today, this level of perfection cannot be guaranteed, particularly for systems of average or high complexity. Also, we are dealing with a situation of great variability, so it is not safe to base decisions purely on, say, "average" process or component defect proportions; they must be based also on possible/credible worst-case situations.

Electronic assembly testers include:

1. Bare-board (PCB) testers, which automatically test PCBs, substrates, wiring harnesses and backplanes for shorts and opens.

2. In-circuit testers, which test individual components (and PCB/substrate tracks) after the components have been placed and soldered. They use a "bed of nails" fixture to connect to the component bond pads. They are very effective at detecting gross component failures (but not all electrical parametric failures on complex devices), and process induced failures such as shorted/open tracks and solder connections, and wrong value or wrongly placed components. It is also useful for diagnosis of the causes of failures detected at later test stages.

3. Functional testers, which test assembled PCBs and higher assemblies via their interface connectors. These testers are expensive, and software must be written for each system to be tested. Modern automatic test equipment often combines in-circuit and functional test.

4. Custom test equipment is often used for particular applications, for example for TV sets, automobile electronic systems and complete military systems.

It is essential that the optimum test strategy is determined for a system, taking account of component and process quality levels expected, test costs, and production throughput. Reference 3 covers this important topic in detail.

Assembly Burn-in

Burn-in of assemblies is applied to find process-induced defects such as damaged components and inadequate soldering, as well as problems due to causes such as component tolerance mismatches. However, it should not be applied as an alternative to component-level burn-in, since the stress levels applied, particularly temperature, are not usually as high (typically not exceeding 80 C for assemblies using epoxy-based printed circuit boards), and also because it is an expensive stage of assembly at which to find defective components.

CONCLUSIONS

Designing and producing reliable electronic systems is a fast-moving and challenging area of modern engineering. Because of the rapid technological changes, in components, their packaging, and in manufacturing and test methods, traditional engineering education has not generally kept pace, so engineers involved in design and production must learn by experience. The very high quality and reliability achieved by the great majority of electronic components and systems shows that the challenges are being met.

REFERENCES

1. O'Connor, P.D.T., Practical Reliability Engineering, 2nd. ed., J. Wiley, 1985.

2. US Military Handbook 338, Electronic Reliability Design Handbook.

3. Davis, B., The Economics of Automatic Testing, McGraw Hill, 1982.

8
■
Failure Mechanisms in Microelectronics Devices

NIHAL SINNADURAI
Consultant, UK

FAILURE MECHANISMS IN SILICON PLANAR DEVICES

INTRODUCTION

The reliability of electronic components is governed by various mechanisms by which they may fail. These mechanisms are dependent on the design, production processes and operational environment of the devices. In order to reduce the incidence of failure it is necessary to determine and study the modes of failure and then feed the information back to the design and production teams for the required corrective action.

In addition to this feedback, it is important that information be supplied to the user concerning the overall reliability of the devices under normal operating conditions. It is, however, often necessary to represent such conditions by some other more convenient reliability test conditions (e.g. thermal stress) and it is consequently essential that these tests be related to operating conditions in a known manner. These latter tests are the accelerated life tests carried out to evaluate device reliability by statistical analyses of the results. The correlation between such tests and operating conditions may be determined by the wearout mechanisms for the devices, where the activation energies for these mechanisms are obtained from a series of step-stress and overstress tests.

Ideally, the device failure rate should be governed by the above mechanisms and give straight lines on lognormal distribution plots, according to an Arrhenius function. Unfortunately such plots are frequently deviated by random or consistent failures and may even be swamped by more dominant failure modes governed by conditions described earlier. In addition, new modes may be introduced by the stress of the life tests. Therefore it is necessary that a 'Physics of Failure' programme determines the failure mechanisms, and their levels of activity. It is common practice to employ screening procedures such as microscopic examination and 'burn-in' programmes, to enable immediate rejection of defective components or to accelerate gross failure potential in a short time.

In this chapter the origins of failure modes and their effects on device characteristics are considered and remedial or compensatory treatments are suggested for them.

The predominant components of today are integrated circuits fabricated in silicon planar technology, whose development and rapid advance has given silicon a clear lead as the technology for Very Large Scale Integration (VLSI), Ultra Large Scale Integration (ULSI) and wafer scale integration. Meanwhile, developments in gallium arsenide technology, with its advantage of a significantly higher electron mobility, promises developments in Very High Speed

Integrated Circuits (VHSIC). Circuits that require special close tolerances or combinations of various monolithic functions are realised by hybrid technologies. This chapter covers reliability aspects of these technologies.

In view of its predominance, more detailed attention is given to silicon planar technology, covering the bulk starting material, the insulating processes, the metallisations, through to interconnections and packaging.

BULK SILICON DEFECTS

To deal with bulk mechanisms, it is first necessary to consider the constituents of the bulk. Modern technology employs a relatively high resistivity epitaxial layer, chemically grown on a wafer obtained by careful cutting and polishing from the original low resistivity crystal ingot. The device junctions are formed within this bulk by carefully controlled high temperature diffusion processes; employing either phosphorus and boron dopants, because of their nearly equal diffusion rates in silicon or by ion implantation of appropriate dopants. It is apparent therefore that the electrical performance of these devices will be governed by the crystal structures and diffusion profiles within them.

Growth·Defects in Crystals

Various growth processes are employed for crystals and these can affect the desired parameters in differing degrees.

Crystal dislocations can result from temperature gradients, plastic deformations and solidification of impurity atoms within the crystal. Flexural strains occurring during grinding and polishing processes employed in crystal slice preparation can also cause dislocation.

Epitaxial layers are also subject to imperfections as a result of their growth processes. These imperfections include dislocations, which are propagated from the substrate where they may result from poor support of the substrate during the growth process.

Imperfections may also originate from scratches on the substrate surface.

The limited solid solubility of the reaction products of some etches and the presence of impurities in the etchants used in substrate preparation before epitaxial growth can result in precipitation on the substrate surface, leading to subsequent faults in the epitaxial layer.

In the future, better control of growth processes will lead to a lower incidence of imperfections in the pre-diffusion crystal material.

Diffusion Induced Defects

The required device topography is achieved by a series of photomasking processes, by means of which windows are cut in diffusion-inhibiting oxide layers. These windows permit diffusions of the necessary dopants to form the device junctions, and to obtain the necessary resistivities of each layer. Diffusion processes are carried out at relatively high temperatures, between

800 C and 1200 C. The choice of dopant, source concentration, diffusion
temperatures and times are all governed by the requirements for resistivity,
junction depth and concentration gradients within the device. The latter design
parameter is important, not for device performance alone, but also in relation
to the surface concentration of dopant, which can influence surface failure
mechanisms as discussed later.

It is important to note here that diffusion profiles are governed not only by
the above process parameters and bulk imperfections discussed before, but also
on the condition of the surface through which the dopants diffuse. Thus surface
irregularities such as erosions, particles and faulty masking will be propagated
as irregularities in the subsequent diffusion profiles.

Various sources of dislocations in crystals have been discussed, but did not
include diffusion-induced dislocations, which outweigh all other sources. There
is, therefore, an inter-relationship in that dislocations cause preferential
diffusion and thus high concentration gradients, while high dopant
concentrations induce dislocations as a result of lattice mismatch.

Diffusion-induced dislocations can be limited by reducing the source
concentration, but a compromise must be reached here with the concentration
required to achieve the desired device performance.

Photomasking irregularities can be minimised by using well defined and correctly
positioned masks and by early renewal of worn masks and fading exposure lamps.
All traces of photoresist should be completely removed and microscopic
examination employed to ensure clean surfaces and well defined windows. As
finer geometries are being introduced with the progress of monolithic integrated
circuits, the optical diffraction limit for photomasking is being reached and
the use of electron beams for finer definitions has been undertaken.

FAILURE MODES ARISING FROM BULK DEFECTS

Anomalous Breakdown

The most significant failure mode resulting from bulk defects is premature
breakdown. The breakdown voltage of a P-N junction is formally defined for an
arbitrary low reverse current level, dependent on the device specification.
Anomalous breakdown occurs by mechanisms which may include the primary mechanism
type, but occurs at lower voltages as a result of fault conditions in the
device. Although failure is formally defined for a particular current level,
the physical onset of failure can occur much earlier and is seen in the shape
and position of the reverse characteristic. The reverse characteristic may be
used to trace the mechanism of failure.

Second Breakdown Mechanisms

Thermal regeneration through local current constrictions (i.e. field enhanced
regions) is instrumental in producing the 'snap back' effects observed in
regenerative second breakdown. Device destruction due to emitter-collector
shorts can occur during regenerative breakdown through local diffusion at the
very high temperatures within the current constrictions.
It is apparent that vulnerability to such modes of failure can be reduced by the
use of homogeneous material and by careful process control. Additionally,
processes may be introduced to counteract current concentrations by: increasing

the collector or emitter series resistance, decreasing the base resistivity and also by modifying the device topography. The latter may be illustrated by the use of linear rather than circular geometry to spread entire region. Increased series resistance increases the forward biased saturation voltage and is unacceptable for switching devices. However, since second breakdown is not a serious problem with switching devices, the cure need not be employed there. Feedback loops can also be grown into devices to stabilise current conditions.

The Reverse Leakage Mode of Failure

In a like manner to the definition for reverse breakdown failure, the reverse leakage failure for a P-N junction is defined for a maximum level at a particular reverse voltage.

Many of the mechanisms for second breakdown also cause high reverse leakages, in particular those mechanisms that provide soft breakdowns. Thus the presence of conducting pipes, dangling bonds, impurity striations and precipitates at the junction will create mechanisms for high leakage. Generation-recombination centres (i.e. trapping sites) in the depletion region are also sources of high leakage dependent on the depletion layer width, which is in turn a function of the applied volts. The effect of inversion at the interface is a major factor contributing to leakage due to the exposure of carriers to a large density of interface states. The resulting larger area depletion layer also exposes carriers to a greater number of bulk generation-recombination centres in the space charge region, with resultant high leakage and lowered emitter efficiency. The various resistive leakage paths provide current levels dependent on the applied voltage. The channelling and tunnelling mechanisms also provide high leakages. It is apparent that the cures applied to the relevant mechanisms for breakdown will also apply to the same mechanisms for leakage.

The hFE Degradation Mode of Failure

hFE is the current gain with the emitter commoned. The specification for transistors normally sets out limits for hFE. The failure criteria normally include these limits and in addition permit a maximum variability (e.g. <10%) in hFE.

The derivations for hFE are dependent on such device parameters as emitter efficiency and base transport factor, which in turn are dependent on the availability of carriers and the generation-recombination centre densities. It is apparent that defects in the bulk provide numerous traps for carriers, thus reducing both the emitter efficiency and base transport factor with resultant decrease in gain. Trapping sites also include crystalline defects such as dislocations (acting as acceptors), polycrystalline inclusions and metallic ions. Neutral impurity atoms can also become traps if they have a low activation energy for dissociation, at the operating temperatures and field conditions.

The efforts to minimise defect occurrences, as described earlier, should be effective in minimising hFE degradation. The use of gettering layers in the insulant layer should also be effective in removing mobile metallic ions from the bulk.
The development of excessive emitter-base leakage by mechanisms described earlier can also degrade hFE, while excess collector-base leakage can enhance

hFE. Either of the two mechanisms is detrimental to stable device performance and must therefore be excluded. The surface channelling mechanism may be prevented by guard ring diffusions. Other mechanisms can be minimised by controlled processing conditions as described before.

INSULATION PROCESS DEFECTS

The major advantage of silicon over other semiconductor materials is the relative ease with which an oxide layer is grown in the silicon surface. In addition to its electrical insulation properties SiO is used as an effective diffusion mask.

In order to evaluate the sources of failure mechanisms it is relevant to consider the various growth techniques and processes for oxide layers.

Pyrolytic deposition techniques employ silicon alkoxy-silanes, deposited on silicon substrates at temperatures as low as 400 C in oxygen or 700 C in vacuum.

Epitaxial techniques employ the reaction of a tetrahalide with water vapour at silicon substrate temperatures of about 1200 C. These deposition techniques grow the oxide on the silicon surface thus leaving active areas of the device exposed to any original surface contaminants and produce a large interfacial state (i.e. unfilled resulting charge site) density resulting from the oxide-silicon mismatch. The techniques have the advantage that the substrate alone is heated and therefore container contaminant diffusion processes are not activated. Pure and controlled dielectrics are obtainable by these techniques.

Thermal oxidation processes employ either steam or dry oxygen. The source gas is passed through a furnace tube containing silicon wafers and controlled at the required temperatures between 1000 and 1200 C. The thermal oxidation process involves reaction of the oxygen atoms with the surface silicon first. Oxygen atoms then diffuse through the top oxide and react successively with the silicon layers beneath, to form new oxide layers. The diffusion of silicon is very small and the excess silicon is concentrated in a very thin layer adjacent to the oxide-silicon interface. The oxidation mechanism also leaves displaced silicon atoms trapped in the oxide. Advantages of the technique are firstly, its simplicity, secondly, that the active areas of the device are removed from the surface by the oxide thickness and thirdly, that the oxidising reaction within the silicon lattice creates minimal interfacial (mismatch) states. The technique has disadvantages in that, at the process temperatures the oxide is contaminated by impurities diffusing through the furnace tubes. Additionally oxide precipitation can occur and induce failure mechanisms.

Dislocation and Diffusion Enhancement

Lattice mismatch can occur with any of the oxidation techniques. For thermal oxides, mismatch is dependent on oxide thickness and oxidation rate. Mismatch can cause dislocation enhancement in the silicon, which is also dependent on the oxide thickness. For oxides thicker than 3000Å, dislocations occur in the bulk (110) planes. A ratio of oxide induced to background dislocations of 50:1 may be obtained for oxides thicker than 4000Å. An effective technique to reduce the of the crystal slice, and to use oxides about 1000Å thick.

Interface States

Interface states are allowed energy levels associated with the termination of the silicon lattice, whose densities are dependent on the oxidation techniques employed. The surface of a crystal is of course a complete mismatch to its surroundings, and therefore has a large trapping state density (surface states). Any techniques that involve growing the oxide on the surface without lattice matching (e.g. deposition) will result in a large interfacial state density. Thermal oxides with the interface away from the original surface and with a relatively good lattice match give fewer interface states.

'Fast states'

This nomenclature arises from the charging and discharging time constants of surface recombination states in the interface region. 'Fast states' are those which have time constants of fractions of a second and are generally located at the interface boundary.

We may consider the interface states to be either donor or acceptor levels, where ionised donors are positively charged, while acceptors are negative. Therefore, the interface state charge can be negative or positive dependent on the surface potential, because it dictates which states are occupied.

As the Fermi level passes through the interface levels, the surface potential changes less rapidly with the applied voltage because the voltage is primarily charging or discharging interface states. This delay gives the effect of 'locking the surface potential.

'Slow states'

Slow states have response times of the order of tens of seconds. They show similarities with certain oxide charge drifts effects but are observed at room temperature instead of the high temperatures required for fixed charge effects. These donor type states probably originate in oxide traps slightly removed from the interface, with time constants dependent on the carrier concentration. Slow states are most pronounced for deposited oxides, therefore the use of thermal oxides reduces their incidence.

Mobile Ions

Charges may also occur as a result of impurities incorporated into the oxide during the growth process. Thermal oxidation processes are carried out at temperatures which activate significant diffusions of contaminants from the furnace linings into the oxidation chamber. Furnace tubes investigated have shown sodium to be the dominant impurity with about 800 p.p.m. for refractory liners, 1800 p.p.m. in Zircon liners and even larger proportions in the various insulating materials used. Quartz boats used for the wafers have less than 20 p.p.m.

In considering the effects of oxide impurities, we should note that surface impurities have a higher mobility than bulk impurities and should be considered separately. The movement of ions affects the surface potential qualitatively if it is lateral over a P-N junction region and quantitatively if it is into oxide.

Oxide Surface Contaminants

Contaminants may be deposited on the oxide surface during oxidation, during subsequent handling processes and from the package ambient.

Residual polar solvents used to remove inorganic residues, photoresist and dust have been investigated by their effects on the resistivity of calibrated distilled water. The investigations also showed that instability can occur on silicon nitride layers because of ionic surface layer formation.

Stability of Passivating Layers

We may now discuss efforts to minimise the presence of contaminant ions in the oxide. The presence of 90% of the bulk sodium contaminant within 200A of the oxide surface suggests the simple remedy of etching away the top oxide layers. The use of thinner oxides will also reduce both the incidence of dislocations and the charge diffusion coefficient, but precautions will have to be taken (i.e. an additional passivating layer) because of the proximity of oxide surface contaminants to any active surface. Removal of 500Å of the oxide surface has been shown to improve oxide stability. The sodium contamination near the oxide-silicon interface (5% bulk) and bulk densities were unchanged by the process, therefore they cannot have contributed to the original oxide instability.

Since a major source of ionic contamination is the hot furnace in which the slices are oxidised, efforts have been directed towards thermal oxidation processes which minimise the diffusion of contaminants through the furnace walls. Two techniques developed have been:

1. A water-cooled cold-walled tube system employing HF heating to avoid heating the quartz tube walls, and
2. A multi-walled hot tube system with a flow of dry gas in between to prevent the diffusion of sodium through the walls into the active area. The latter process employs either an independent flow of dry gas or the outflow of the oxygen through the outer walls.

In addition to the above, metal sources should be degassed in vacuum prior to contact evaporation.

Phosphorus Glass

Phosphorus is used as a dopant in the oxide because it acts as a getter of sodium.

The advantages of phosphorus diffusions led to the development of phosphosilicate glass layers as gettering layers for alkali ions. The technique involves an overlay of phosphosilicate glass on the silicon dioxide, achieved by diffusion of P_2O_5 into the SiO_2. The diffusivity of P_2O_5 in SiO_2 is fast and drops off sharply as its concentration drops below that of the glass, thus giving a nearly flat profile which drops off steeply. The transition from phosphosilicate glass to undoped SiO extends over only 100-200Å. Borosilicate layers also have gettering properties for alkali ions and may be employed in a similar manner.

Silicon Nitride Passivation

The vulnerability of oxide layers to ionic contamination led to consideration of alternative passivating layers, resulting in the development of silicon nitride films. Silicon nitride processing has a disadvantage over oxide processing because it is chemically grown or deposited.

Nitride deposition techniques unfortunately result in high densities of interfacial states. Therefore Si_3N_4 processes normally include the growth of a thin (100-200Å) layer of thermal SiO_2 prior to the deposition of 1000Å of Si_3N_4. The complexity of the nitride passivation process and subsequent masking processes is one of its great disadvantages.

The advantages of Si_3N_4 over SiO_2 are:

1. Higher breakdown strength of 10^7 V/cm ($SiO_2 = 6x\ 10^6$ V/cm)
2. Higher dielectric constant ($Si_3N_4 = 9$ $SiO_2 = 3.8$)
3. Denser and therefore better diffusion barrier
4. Greater passivity

The Ideal Passivation

From the foregoing considerations, the optimum passivating layer may be obtained as follows:

1. Thermally grow a silicon dioxide layer, of about 400Å thickness, on a wafer previously etched on both sides, employing cold-wall or multi-wall tubes.
2. Carry out a subsequent phosphorus diffusion into the oxide, using sufficient source concentration to obtain a phosphorus glass-like profile.
3. Etch away the top 200Å of the oxide to remove the excess surface concentration of the Na contamination.
4. Grow a 1000Å Si_3N_4 layer over the residual SiO_2 layer.

FAILURE MODES ARISING FROM INSULATION DEFECTS

M.O.S. Surface Potential Shifts

The effects of interface states and charges in the oxide, causing shifts in the C-V curves of M.O.S. transistors, have been discussed. The origins of these shifts have also been discussed. The generally observed N-type tendency of the Si surface shifts the 'turn on' voltage of N channel M.O.S.T.'s such that the channel has to be turned off by a negative gate bias.

Any drifts with time represent instabilities in the operating characteristics of M.O.S. devices and thus unreliability. Although stability with time can be achieved by the use of passivating layers described, the shifts from the ideal situation will remain.

Depletion and Inversion Induced Modes in Bipolar Transistors

The mechanism of oxide charge drift also causes instabilities in bipolar transistors, the most significant effect being the depletion and inversion of the surface. The condition may be encountered after reverse bias and high temperature stresses, when charges in the oxide may be swept directly above those silicon regions which are vulnerable to depletion or inversion. The presence of positive ions is more common, but negative ions have been observed to cause inversion of the less highly doped collector N-type layers of N-P-N transistors.

hFE mode surface depletion of the base near the emitter-base junction causes the base current to increase to accommodate recombination losses due to enhancement of surface recombination as well as bulk recombination in the surface depletion region, while the collector current is unaffected, thus causing a decrease in hFE.

'Channel' leakage gives a resistive path from emtter to the base contact or excessive leakage by recombination effects at the unpassivated chip edge. It should be noted that P-N-P devices are more easily inverted because of the lower doping level in the collector region.

We may consider the effect of a channel as follows: at reverse bias, current flows along the channel to the base contact, producing a lateral voltage drop in the channel, causing 'pinch off', and giving a channel saturation current independent of V. The pinch off voltage is dependent on the induced channel and the bulk dopant levels and saturation may not occur before the undegraded VBR is reached.

Channel or surface leakage can cause changes in hFE. Collector to base leakage will cause apparent increases in hFE for low currents of the same order of magnitude as the leakage current. Emitter to base leakage on the other hand causes an apparent decrease in hFE at low currents. Both effects are significant for low currents but become insignificant at higher currents.

Inversion effects may be annealed out with zero bias at high temperature.

Channel effects can be inhibited by the use of a guard ring or a channel stopping diffusion. Guard rings are excess P-type (P+) diffusion rings, in P-type layers, which encircle N-type regions. They stop the channel spreading by making inversion more difficult.

Recombination losses can still occur if there are surface defects within the guard ring. The additional safeguards are essential to obtain reliable P-N-P transistors.

Surface leakage through process contaminants on the surface is reduced to a minimum by the use of double distilled, deionised water as the final cleaning solvent.

METALLISATION AND INTERCONNECTION DEFECTS

Once the basic device has been fabricated, it is necessary that ohmic contacts be made to the various active regions. Such contacts are achieved with metallic layers, formed either by evaporating, or by depositing from a glow discharge, into windows cut in the passivating layers. For M.O.S. devices the metallic gate is deposited on the oxide. Alternatively polysilicon may be used.

Certain devices such as 'space charge limited' devices and 'metal base' transistors employ metals with a high metal-to-semiconductor work function difference (ΦF) (i.e. non ohmic). However, for most general applications, especially with silicon technology, a low ΦF metal such as aluminium is used. Gold, silver and various compound metallisations have also been employed.

Aluminium Metallisation and Associated Faults

The most commonly encountered metallisation with silicon planar technology is aluminium which has the properties of good adhesion to silicon, a low metal to semiconductor work function difference, a relatively high electrical and thermal conductivity and a relative ease of evaporation for processing. Contact to the active areas of devices is usually achieved by depositing a controlled thickness of aluminium through windows cut in the passivating layer. Interconnections on interdigitated structures and integrated circuits are made by masking and then etching the required aluminium patterns.

Pinholes and Other Process Faults

The effects of particle contamination and faulty masking can cause defects. Faulty etching of the aluminium can also result in irregular metallisations.

Mask misalignment is another fault encountered occasionally and results in contacts which do not wholly occupy the contact window. Potential failure mechanisms are therefore present in that the actual contact has to carry a higher current density; secondly, a non-uniform current distribution exists in the silicon window and thirdly the contact area, although protected by a thin oxide, is vulnerable to contamination.

Handling damage may also result in cuts and smears, which are potential failure regions. Therefore careful handling is essential. The masking processes must be carried out in clean room conditions to ensure the contamination does not cause pinholes or voids.

Aluminium Spikes and Irregularities

Faults in the material also lower the activation energy for the movement of aluminium under high field conditions. Aluminium tracking can occur between adjacent contacts following destructive breakdown at the surface. In certain instances the tracking may be observed through a microscope to commence as bias is applied and cease when the bias is removed. Tracking has been observed at both inputs and outputs of integrated circuits. The common cause has been attributed to an over volt spike which causes a discharge phenomenon either over or under the oxide, dependent on whether the metallisation was overlaying, or wholly within, the contact window. The discharge commonly carries metal particles with it, thus causing a tracked leakage path, which alloys into the silicon when under the oxide.

Corrosion

Aluminium, as opposed to gold, is prone to corrosion effects, especially in the presence of chlorine. The reaction of chlorine with aluminium is governed by the presence of water vapour:

$$6 \; HCl + 2 \; Al \longrightarrow 2 \; Al \; Cl_3 + 3 \; H_2$$
$$2 \; Al \; Cl_3 + 6 \; H.OH \longrightarrow 2 \; Al \; (OH)_3 + 6 \; HCl$$
$$2 \; Al \; (OH)_3 \; \text{Ageing} \longrightarrow Al_2O_3 + 3 \; H_2O$$

which is a cyclic sequence dependent on the availability of moisture.

To limit corrosion reactions it is necessary to control the package atmosphere and exclude both chlorine and moisture. Chlorine usually arises from the breakdown of residual solvents, but may also originate from the package materials. The suggested use of double distilled deionised water as a final solvent should be effective in cleaning previous solvents, while package sourced chlorine can be eliminated by replacing the package type. Moisture content can be limited by controlling the final encapsulation stage. The deposition of an amorphous oxide or nitride over the aluminium inhibits corrosion.

Aluminium Drift or Electromigration

Extensive studies have been made of the movement of aluminium in IC conductors for highly polycrystalline (deposited on to cold substrate), well ordered (deposited onto hot substrate) and glass protected well ordered aluminium films. For each type, observations were made of the mean time to failure (MTTF) of the conductors and, in addition, the movement of razor edge notches in the conductors were studied. These notches behaved as vacancies and were observed to drift opposite to the electron flow, while aluminium was observed to aggregate in the direction of the electron flow. Two alternative mechanisms were proposed for the lifting of aluminium atoms out of the lattice: first, that the positively ionised aluminium atoms would tend to drift towards the negative contact end; secondly, momentum exchange with electrons would cause aluminium atoms to drift towards the positive contact. The observations showed that the latter mechanism dominated for current densities of the order of 10^9 A.m^{-2}. An equation for the mechanism was derived, which gave a relationship between the MTTF and current density (J) according to:

$$MTTF = 1/A \times 1/J^2 \cdot \exp(Ea/kT)$$

where A is a constant, Ea is the activation energy and T is absolute temperature. The value of Ea is typically 0.48 eV for polycrystalline film, 0.8 eV for well ordered film and 1.2 eV for well ordered film with glass overlay. The latter value approaches the self-diffusion activation energy for bulk aluminium. Therefore, the lower activation energies for the unprotected films can be accounted for by diffusion along grain boundaries and surface diffusion. At 270 C the ordering of the films is immaterial and the activation plots for the polycrystalline and well ordered films intersect. Investigations of the emitter conductors of working devices have agreed closely with the observations for polycrystalline films and gave activation energies of about 0.45 eV.

That the drifts are related to dc and not just to power dissipation in the conductors has been confirmed because a.c. dissipation resulted in much longer lifetimes than for d.c. The lifetimes were also seen to decrease with increases in current densities greater than 5 x 10^9 A.m^{-2} and also with increases in temperature. Correlations have also been established between electromigration effects and the bulk self-diffusion coefficients for a number of different metal films.

Electromigration effects can be minimised by limiting the current densities to less than 10^9 A.m^{-2}, e.g., by the use of interdigitated structures.

Geometry

It has been shown that localised hot spots cause degradation of devices. These hot spots may be detected by the use of an infra-red microscope and usually occur (in metallisations) over active junctions, at oxide steps at contact cuts and constrictions in the aluminium. Localised breakdown also causes hot spots. significant in that current tends to crowd towards the inside of a corner with a resultant higher current density, causing localised heating. Such heating would only be significant with thin and narrow metallisations, such as those employed in the latest VLSI circuits. Therefore the design of such metallisations should exclude sharp corners. Constrictions at oxide steps and contact cuts may be avoided by using multiple coil evaporators to ensure uniform deposition.

Alternative Metallisations

Although aluminium is the most widely used metallisation, other metallisation systems have been considered, to overcome some of the disadvantages of aluminium. In considering these metallisations we must also pay regard to the type of bonds employed.

Gold metallisations do not adhere sufficiently strongly to silicon and do not reduce the residual oxide in the contact area to give good ohmic contact. Furthermore, gold atoms readily diffuse into the silicon with resultant degradation of the electrical properties of the device. Therefore two and three layer contact systems have been developed including gold chromium and gold molybdenum, the second metal of each pair being used to provide a low resistance contact while the current is carried mainly by the gold layer. However, Au-Mo systems have been found to fail after stress tests due to the peeling of gold from the molybdenum. Corrosion of the molybdenum may also occur. The most widely used three layer gold metallisation is titanium-platinum-gold.

Bonding

The IC metallisations and package pins are usually interconnected by fine (0.001 i dia.) wires which are bonded to the leads and to special bonding pads on the metallisations. The wires chosen for such bonds must be compatible with the metallisation and packaging systems used. To date bonding wires have been mainly either of gold or aluminium.

Thermo-compression and ultrasonic bonding are still predominantly employed in device production. The techniques, as the names imply, require the application of pressure or ultrasonic energy from a bonding tool onto the wire and contact pad while the device is heated. The structure of the bond depends on the type of tool employed. These may be either chisel bonds or ball bonds.

Gold wire is widely used and has been bonded, employing chisel and ball bonds, onto aluminium metallisations for many years. Both gold and aluminium wires are compatible with the conventional TO-type package leads, which are usually of gold plated kovar for IC packages.

Unfortunately gold bonds on aluminium metallisations are subject to certain degradation phenomena, both chemical and physical in nature.

The disappearance of aluminium metallisations adjacent to gold bonds has been attributed to the self diffusion of aluminium atoms towards the bond area, which acts as a sink for the aluminium. The mechanism also results in loss of adhesion between the aluminium and silicon. The migration of aluminium may be inhibited by depositing an amorphous oxide passivation layer over the wafer.

Intermetallic Compounds

A common problem encountered with bonds employing different metals has been the formation of thermally activated intermetallic compounds, which may degrade the contact properties. The name 'purple plague' has been frequently used to describe the undesirable compound formed by the reaction between gold and aluminium bonds. However, investigations have shown that the purple (aluminium rich) compound (AuAl) far from being weak is both mechanically strong and highly conductive electrically. Instead it appears that a tan phase (Au Al) compound which develops with silicon as a catalyst, is the brittle plague which gives poor electrical conductivity. A black, gold rich plague has also been observed, which results from the diffusion of gold into the surrounding aluminium, and gradually spreads through the metallisation. Black plague formation causes voids in the adjoining aluminium. With expanded contacts it is quite feasible that void formation around the bond will isolate it from the contact pad with a resultant open circuit. This 'Kirkendall' effect has been observed to result in open circuits. Microcracks occur at the periphery of the bonds because of insufficient interdiffusion of the metals in the last few milliseconds of cooling. These cracks are widened into voids during black plague formation. Strain resulting from the crystallographic mismatch between the different intermetallic phases may also cause strains in the materials.

As heat is an activator of Au-Al intermetallic growth, aluminium wires are bonded to aluminium metallisations on ICs packaged in ceramic packages processed at high temperatures, whereas the robustness of gold wires is preferred for the low temperature high quantity automated manufacturing processes for plastic packaged ICs.

Aluminium wires on gold metallisations have more disadvantages than gold on aluminium at the chip and in addition have a potential for plague formation. Replacement of aluminium as the metallisation is impractical, because of the good adhesion properties of aluminium, therefore an Al-Al system would be most advantageous. The catalysing effect of the silicon can be avoided by removing the silicon.

Bond Orientation

The orientation of the bonding wire should be considered in relation to the stresses undergone by the devices. Slack wires may sag during mechanical shock and cause short circuits to other contacts. Taut wires cause mechanical strain at the bonds with resulting deterioration. Excess wire beyond the bond (long tails) with chisel bonds may bridge the gap to an adjacent contact and cause shorts. It is also essential that the bond is centrally positioned to ensure full use of the contact area. The bonding parameters (pressure/energy, temperature and time) should be chosen to ensure that underbonding (resulting in poor adhesion) or overbonding (resulting in thin wires with a potential for open circuits) does not occur.

Tape Automated Bonding (TAB)

Tape automated bonding is a simultaneous-bonding alternative to wirebonding and packaging and provides robust tape fingers bonded directly from the IC chip to the eventual circuit. The interface contact metal is usually gold for thermosonic bondability, while the outer termination is usually soldered. TAB is normally plastic encapsulated, but may be ceramic sealed.

Package Types

'Dual-in-line' packages (DIP) are so described because they have two sets of in-line terminals, one on each side of the pack. They have similar advantages and constituents as 'flat packs' but are larger and sturdier. For the ceramic hermetic package ('CerDIP') the materials used are ceramic sandwiching a solder glass, which is used to achieve a relatively low temperature hermetic seal, and gold plated kovar leads. Moulded plastic packages ('PDIPs') are those in which a plastic such as an epoxy, phenolic or silicone is injection or transfer moulded around a chip already mounted and bonded to a lead frame. The technique lends itself readily to automation and hence to lower costs for mass-produced devices.

'Flat packs' are small flat packages made up of combinations of ceramic, metal and glass and more recently of plastic. Earlier packs had fragile leads, but they have now been made sturdier.

Chip carriers have leads on all four sides and are much more compact than DIPs. They are mounted to the surface of the circuit substrate, rather than through holes in a printed circuit board. (Hence the name 'surface mounted device' SMD). The main reliability problem is the integrity of the solder joints.

Another type of SMD is the pin grid array, which has an array of connections on the underside.

SMDs are becoming standard for many applications, but the solder process requires close control to ensure integrity.

Properties of Plastic Encapsulating Materials

Analysis of failure generated during accelerated ageing tests have shown that certain properties of the plastics encapsulating materials contribute to component failure.

From the observations of corrosion and parametric degradation, it was evident that the leaching of acidic or alkaline substances or ionic contaminants from the plastic, particularly during moisture ingress, were distinct hazards. Consequently, suitable limits were set for the pH and conductivity of the water-extractable substances from the plastic. It is also necessary to avoid damage to the bond wires of the ICs by ensuring either that the coefficient of expansion of the encapsulant is nearly matched to that of the bond wires, or that the encapsulant is sufficiently flexible to avoid exerting excessive shear or tensible forces on the wires. In order to safeguard the insulation resistance between the terminations and thick-film conductors, it is also necessary to limit the permitted silver reducing power of the plastic. When considering the valid range over which thermal overstress testing may be carried out, an important ceiling is the glass transition temperature, which should therefore be determined. These various requirements have been embodied in procurement specifications for type-approval of plastic encapsulating materials and typical values are given in Table 1.

Table 1

Acceptance Limits for Some of the Material Properties of Plastic Encapsulants

Property	Limits
Water extract pH	4 - 9
Water Extract Conductivity	<0.03 S/m
Coefficient of Linear Thermal Expansion	<35 x 10^{-6}/°C (or flexible)
Flexibility	Bend over 6 mm Mandrel
Silver Reducing Power	<5 mg/g
Glass Transition Temperature	Measure

The reliability of hermetic and non-hermetic packages is covered in Chapter 7.

FAILURE MECHANISMS OF GALLIUM ARSENIDE DEVICES

Introduction

GaAs offers significant potential advantages over silicon, with a higher electron mobility at low fields, a negative resistance in the I/V characteristic, and the ability for integration with other III-V compounds. These benefits have been translated into actual device realisations - initially IMPATT diodes and transferred electron devices (TEDs), and then with short, (0.25 micron) gate FETs for low noise amplifiers and power amplifiers leading to monolithic microwave ICs (MMICs). However, reliability achievements have been somewhat slower.

Bulk Mechanisms

Many of the failure mechanisms in GaAs devices are similar to those of silicon devices. In addition, there are other failure mechanisms related to GaAs. Failure mechanisms originating in the bulk material, due to GaAs substrate defects can affect both yield and performance. Dislocation densities are in the range $10^5 cm^{-2}$, so that it is more difficult to produce large area GaAs circuits.

Metallisation and Interconnection Mechanisms

Metallisation and contact failure mechanisms are similar to those in silicon devices, but are initiated in different ways. Such mechanisms occurring in GaAs FETs are summarised below:

Mechanism	Cause
Electromigration of Ohmic contact	Ga outdiffusion
Open circuit in the gate	Al burnout

In GaAs, electromigration is promoted by an initial out-diffusion of Ga which results in a Au-Ga top surface. Electromigration can be inhibited, for instance, by doping the gold contact metal or by introducing diffusion barriers. The activation energy for electromigration of gold on GaAs FETs is about 0.5 eV, indicating that low-temperature grain boundary diffusion is the predominant mechanism. Electromigration also occurs in power FETs which use AuGeNi for the low ohmic contacts.

Another mechanism of significance is the occurrence of voids in the gate metallisation arising from Kirkendall interdiffusion effects causing the gate to disappear from the periphery.

In the higher frequency (up to 170 GHz) microwave devices, operation is at much higher temperatures, typically 180-240 C for IMPATT diodes. At such high temperatures, intermetallic reactions are promoted more rapidly, resulting in critical onsets of runaway. Particularly with IMPATT diodes, which operate in the avalanche mode, the smallest imbalance in uniformity of avalanche current leads to higher local temperatures causing extra electron-hole pair generation locally thus resulting in further increased temperature, thermal runaway and molten destruction of the diode. Studies of IMPATT diode reliability have confirmed a close link between the activation energy for intermetallic reaction and failure of IMPATT diodes. Reliability improvements can be achieved by designing more stable metallisation systems, and by good thermal designs to reduce the thermal resistance, e.g. by creating an equivalent array of diodes.

FAILURE MECHANISMS OF HYBRID MICROCIRCUITS

Bare-Chip-and-Wire

Hybrid microcircuits comprise thick or thin film resistors and interconnections, and other added components including diodes, transistors and integrated circuits. The semiconductor components might be pre-packaged, in which case their reliability would be as described earlier. However, the demand for high packing densities led to use of the semiconductor devices in bare-chip form, wire-bonded directly to the hybrid substrate. As hermeticity was regarded as essential to acheive high reliability, new and (large) hermetic packages were developed to contain the entire hybrid substrates. With few expensive exceptions , the bar chips as well as the ceramic substrates are glued down with organic adhesives within the hermetic package. Solder adhesion is used for high-reliability packages.

These new technologies were adapted and taken for granted with very little understanding or evidence of reliability performance. The new environment within the encapsulation was in fact likely to be different from that normally encountered by the individual ICs and thick-film networks. Nevertheless, provided that no new failure mechanisms were induced in either the ICs or the films, then conventional thermal overstress could be regarded as a relevant reliability assessment test. However, reliability studies of the bare-chip-and-wire (BCW) technology revealed that a new hazard was indeed introduced by the new technology: namely that the organic adhesives outgassed ammonia and/or water vapour. Also, as the vapours were trapped within the hermetic enclosure, they were concentrated around the very components that were supposed to be protected by the package. The rate of outgassing has been found to be accelerated by temperature and to conform to the Arrhenius expression. Consequently, it is possible to assess reliability by carrying out thermal overstress, which has revealed very poor reliability indeed due to the use of amine-cured adhesives. The significant outgassing of vapours revealed by such overstress, and expected during the operation of such hybrids, was confirmed by analysis before and after overstress. The elimination of ammonia did not achieve a great improvement in reliability because alternative adhesives evolved more water vapour instead. However, complete elimination of the substrate adhesive, by integrating the substrate with the package base in the Integral Substrate Package (ISP), did achieve a marked improvement in reliability.

Thick-Film Resistors (TFRs)

Detailed analyses of progressively ageing resistors showed that the mechanism of drift is consistent with tunnelling and trapping of carriers at sites associated with a narrow band in which conduction occurred. Thus the basic mechanism of loss or recovery of conductivity in resistors may be regarded as akin to the recombination effects on the gain of transistors.

Many such effects can be effectively inhibited by the use of an overglaze to protect the TFRs. Other instabilities in TFRs have been identified to be due to mechanical stress or flexure of the substrates.

Whilst screening tests can help avoid the procurement of unstable components, thick-film circuits should be manufactured in a manner suited to achieving high reliability. Examples of procedures that can be used to achieve higher reliabilities from thick-film resistors and conductors include:

1. Overglazing with a compatible glaze.
2. Avoiding an overall plastic encapsulation.
3. Locating sensitive resistor elements away from high voltage elements.
4. Ensuring symmetry of geometry and electrical layout of resistors that are required to track closely.
5. Designing the size and trim of close tolerant resistors to limit power dissipation to less than 60 mW/(mm^2) .

REFERENCE

Sinnadurai F. N., (ed), Handbook of Microelectronics Packaging and Interconnection Technologies, Electrochemical Publications, Scotland, 1985

9

Reliability Assessment of Microelectronics Devices

NIHAL SINNADURAI
Consultant, UK

INTRODUCTION

Methods of assessing component reliability have received considerable attention by military, telecommunications and computer administrations over the past two decades and, with the support of considerable experimental evidence, have become widely accepted. Reliability assessment, however, is not and is unlikely to become an exact science.

The Bathtub Curve and Its Reliability Regimes

The reliability behaviour of microelectronics components can affect different stages in the life of electronic equipment. These different regimes are usually characterised by the classic 'Bathtub' curve (Fig 1) in which the early sharply decreasing failure rate is associated with a period of infant mortality or de-bugging of manufacturing faults in components and assembly processes, the constant failure rate in mid-term occurs because of random defects in use, and the rising failure rate at the end of life indicates component wear-out. However, in a properly controlled production line for reliable components it is unlikely that early faulty manufacturing procedures, giving rise to unacceptable high infant mortalities, would persist. For non-hermetic (plastic encapsulated) ICs the more usual occurrence is ultimate wear-out, as verified by many endurance and accelerated ageing tests on a wide variety of component types. However, this is a problem which has been greatly reduced by modern protection and packaging methods, and is only likely to occur in severe environments over long periods.

Infant Mortality and Burn-in

Burn-in has become an accepted practice as a means of accelerating infant mortality in a population of semiconductor devices. Both manufacturers and users apply burn-in when appropriate. As proper burn-in requires that the devices under test (DUTs) be plugged into parallel sockets for periods ranging from 4 to 168 hours, a demand has arisen for equipment capable of performing some kind of functional test during the long period in which the devices are in 'captivity' during the burn-in cycle. This demand has intensified as memories and logic devices have become more complex, lengthening test times to unacceptable extents. Today, networked burn-in/test systems are acapable of fully exercising complex ICs during the burn-in cycle, and of computing the hazard rate against a target rate and terminating burn-in at the optimum point.

Extracts from "Handbook of Microelectronics Packaging and Interconnection Technologies" published with the kind permission of Electrochemical Publications Ltd, 8 Barnstreet, Scotland KA7 1XA.

As a general rule it is best to burn-in at the component level when the cost of testing and replacing defects is lowest. Components can be stressed to their limits, resulting in shorter burn-in times. Burn-in of a board or an assembly is difficult, and higher temperature devices will not be stressed enough. ICs need specific bias conditions to trigger certain failure modes, a condition that cannot be applied to all components at the board assembly level. However, burning-in of boards or assemblies can show up assembly faults, such as cold or dry solder joints and contact problems.

Testing during burn-in saves manufacturing time. However, it costs more because of the complexity of the equipment necessary. One of the most expensive parts of burn-in is the physical loading and unloading of large numbers of devices in boards. Here automatic loaders and unloaders can be a worthwhile investment.

Static burn-in is typically performed on discrete semiconductor devices and SSI/MSI for 48 - 168 hours for simple parts or test times appropriate to the potential application. Dynamic burn-in is used for the more complex devices, such as microprocessors and memory products, where thermal stress combines with dynamic stimulation of inputs to provide worst case operating conditions. The principal argument for static burn-in is that it is necessary to induce contamination-related failure mechanisms where a steady-state bias (voltage field) must be maintained at high temperature in order to accelerate the migration of impurities to the surface so that failures will occur during the burn-in cycle. The advantages of static burn-in include low system costs and simple programming. It is also effective in accelerating contamination, surface-charge separation and inversion failure mechanisms. The principal disadvantage of static burn-in is the fact that it is 20 - 40% less effective than dynamic burn-in for LSI and VLSI devices. It is for this reason that most LSI/VLSI devices such as memories and microprocessors are burned-in dynamically in order to stress as much of the internal circuitry as possible. Dynamic burn-in has the principal advantage of being much more effective, particularly for LSI and VLSI devices, because it stresses all semiconductor nodes, dielectrics and conductive paths electrically. Disadvantages of dynamic burn-in systems include much higher equipment costs and the requirements for more dedicated burn-in boards.

Test times for complex semiconductor devices have become unacceptably long. The time required to run a simple N pattern such as GALPAT on a typical 16K RAM is one minute; for a 64K RAM, it increases to 30 minutes; and for a 256K RAM, it becomes eight hours. Even simpler patterns have become uneconomically time consuming for large memories when run on a conventional single-head tester.

At the same time, there has been a steady decline in device prices and a steady increase in the cost of testers - with no significant increase in throughput. Each device must still be tested sequentially, although some test systems test two or four devices in parallel.

Since burn-in requires that thousands of devices reside in sockets for many hours, it is economically desirable to shift some functional testing to the burn-in chamber. There is ample time during burn-in to run repeated pattern tests that will exercise even the largest memories available today through a wide range of test conditions.

Improving Burn-In Throughput

New, state-of-the-art integrated circuits are characterised, typically, by low yields and high test costs. During the early portion of the IC product life cycle, product engineering strives to improve the yield, while production strives to reduce the cost of final test.

By monitoring outputs during the test/burn-in cycle, the network can calculate the instantaneous hazard rate and terminate burn-in when the failure-rate curve begins to flatten out. As this target rate may vary substantially from lot to lot, automatic termination can save substantial time and test cost.

The cost of testing has increased, both in absolute terms and as a percentage of total manufacturing cost. At 1982 rates, packaged-device test cost about $2 per minute, or about $0.75 each for the 64K DRAM in the early portion of its product life cycle. Notwithstanding changes in test strategies, this forbodes even higher packaged test costs for the 256K and 1M DRAMs in the early portions of their life cycles. Test during burn-in can help to reduce these costs.

Test during burn-in can be used to detect early failure mechanisms and to reduce burn-in duration through iterative design and process improvement. Using test during burn-in, it is possible to log failures during the accelerated life test without having to remove the product from the thermal chamber. The process not only can reduce the early failure rate by 50%, but also can reduce the target burn-in time from about 96 hours to less than 24 hours.

In addition to infant mortalities that may respond to acceleration by increased stress of some form, there are those that may be inherent in the material chosen, e.g. soft errors sourced from the package material, or as a result of inadequate precautions in handling resulting in electrostatic damage (ESD)

Soft-Error Screening

For VLSI memories, detection of 'soft' errors has become extremely important. The high density of these memories makes them susceptible to transient ('soft') errors resulting from bombardment by alpha particles. However, 'soft' errors occur at rates that are measured in terms of errors per million device hours. Thus, to determine soft-error rate, it would be prohibitively expensive to accumulate sufficient test data on conventional testers. For example, one would

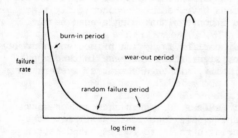

Classic bathtub reliability curve.

Figure 1

134

have to exercise ten devices for 100,000 hours to achieve the equivalent of a million-hour sample. However, with a network of just four system test assemblies, each containing 8000 DRAMs, alpha particle characterizations can be performed in as little as 30 hours. Moreover, when such a system is used for 100% production burn-in, the network generates alpha-particle data as a by-product of testing during burn-in.

Alpha-particle-error analysis requires more than just the ability to detect failing devices, which early test/burn-in systems provided. It also requires the detection of failing bits. Once a failing bit has been detected, a sophisticated analysis algorithm may be performed by the network controller, which causes a different testing sequence to be run. This second sequence distinguishes between hard errors, intermittent errors, and genuine alpha-particle failures. A report can then be generated that lists the addresses of failing bits.

Electrostatic Damage

Detailed experimental attention has been given to the simulation of the problem of ESD to components, before and after assembly in equipment. Static charging and discharging can occur during all stages of component and equipment production, assembly, test and field use. Therefore the use of adequate ESD precautions can be a major step in reducing infant mortality or early life failures.

Wearout Failures

There are two approaches that are commonly adopted in assessing wearout reliability. One is to determine the criteria for acceptance or rejection of certain properties of the components that are known to contribute to their unreliability (e.g. hermeticity). Moreover, as the reliability of electronic components is governed by the mechanisms by which they may fail, the specification of tolerable limits has necessitated the study of failure mechanisms. The use of tolerable limits or otherwise screening components is sometimes referred to as Reliability Indicator methods. The other approach is to determine whether the components will survive the desired duration by accelerating the failure mechanisms in a predictable way.

The desired longevity of electronic components for the higher reliability applications is usually about 20 years (175,000 hours). The problem that faced the reliability engineer was how to simulate such long lifetimes in much shorter time scales that would be acceptable when routinely assessing component reliability. Thus, time compressions of the order of 200 or more were sought, requiring accelerated ageing tests to be devised. The studies of accelerated ageing led to a better understanding of the tolerable values of initial properties and thus also aided the specification of acceptancce criteria where appropriate.

ACCELERATED AGEING

The ageing process of a packaged component is very dependent on the sensitivity of that component to the environment permitted by the package to reach the component. For instance, a truly hermetic package would prevent deleterious gases and vapours from reaching the component within, and consequently the

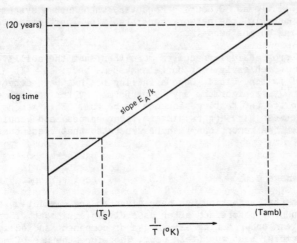

The use of the Arrhenius Plot to select thermal overstress temperatures and tin simulate 20 years operation at ambient temperatures (Known E_A).

Figure 2

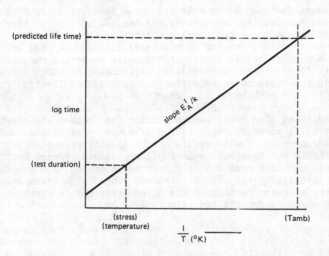

The use of the Arrhenius Plot to predict reliability from results obtained from thermal overstress tests (Unknown E_A').

Figure 3

ageing of the component would be governed only by the electrical and thermal stresses inflicted on it. These various influences are considered below.

Accelerated Ageing by Thermal Overstress

The classic relationship that describes reaction kinetics is the Arrhenius equation, which relates the rate of reaction (R_T) and the absolute temperature (T) in the following manner:

$$R_T = const \times exp(-Ea/kT) \tag{1}$$

where k is Boltzmann's constant and Ea is referred to as the activation energy. Extensive reliability studies have confirmed that the Arrhenius relationship does indeed apply to the thermal ageing processes of semiconductor components, and has led to the concept of accelerated ageing by thermal overstress, in which long term operation is simulated by exposing components to elevated temperatures for relatively short durations dependent on the acceleration factor achieved by the overstress. The manner in which time compression is achieved is shown by rewriting equation (1) thus;

$$log\ t = const + Ea/kT \tag{2}$$

hence log t is directly proportional to the reciprocal of the absolute temperature, as shown in Figure 2. Therefore, if Ea is known, then the longevity t1 at a temperature T1 may be related to the longevity t2 at a different temperature T2, as shown in Figure 2 or as follows:

$$A_T = t1/t2 = exp\ Ea/k(\frac{1}{T1} - \frac{1}{T2})$$

in which A is the acceleration factor for thermally accelerated ageing, and is a measure of the time compression achieved by raising the component temperature to simulate operation at the lower temperature. Thus, if the desired lifetime is, say, 20 years at an operating ambient temperature of, say 70 C then the Arrhenius line may be anchored at the intersection of these two parameters, and extrapolated to the equivalent test condition at a higher temperature and shorter duration, which is the basis of thermally accelerated ageing testing.

If, however, Ea is unknown then component failures must be induced at a number of controlled conditions, and the times to failure (tn) related to the test temperatures (Tn) for a particular percentage of component failures. If, for instance, 10% is regarded as the tolerable number of failures, the test would be conducted until sufficient failures occurred to provide confidence in the times to 10% failures at each of a number of elevated temperatures, and a straight line through the points log tn, 1/Tn aould have a slope of Ea/k and could be extrapolated to the acctual operating ambient temperature T to determine the actual lifetime that would be obtained (Figure 3).

The real situation is not quite so straightforward, however, because genuine acceleration of ageing requires that the whole distribution of component failures should be consistently advanced with each step of increased thermal stress. Consequently, it is necessary to obtain a significant part of the time-dependent failure distribution at each temperature in order to determine whether the distributions are actually shifted monotonically with the reciprocal of the absolute temperature. In practice, it has been found that many electronic components fail according to either log-normal or Weibull distributions, and plots of cumulative failures versus log time can yield straight lines on log-

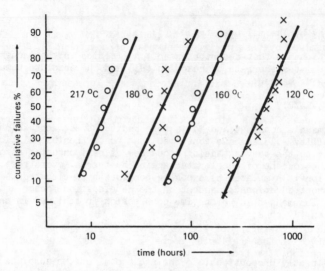

Log-normal failure distribution accelerated by temperature.

Figure 4

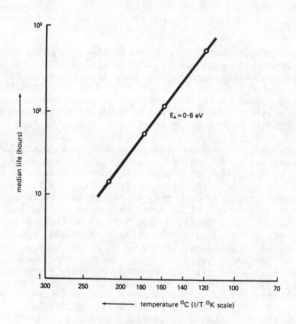

Arrhenius Plot of median life vs temperature (From Fig. 4).

Figure 5

normal graph paper (Figure 4). The intercepts at 50% failures, i.e. 'Median Life', may be plotted to give an Arrhenius plot (Figure 5), which strictly should be obtained from regression analysis of all the failure data obtained, to produce the best fit straight line together with appropriate confidence bands (Figure 5) which reveal the uncertainty of the extrapolations. Increased confidence is obtained by testing larger numbers of components. Component failures do not always conform to log-normal or Weibull distributions which may be wrongly impressed on the failures that are obtained, and should be justified by regression analysis of the data. Another necessary precaution is to extrapolate the Arrhenius line only up to temperatures that may be validly employed to accelerate ageing. It is possible, for instance, that electronic components simply cease to function or that abnormal failure mechanisms are induced above (or below) particular temperatures. These 'validation limits'should be determined as part of any programme of reliability evaluation of components. Examples of validation limits are given in the following sections.

Accelerated Ageing by Damp Heat Overstress

As the use of plastic packaging of ICs became widespread and made inroads into the former preserves of the 'reliable' hermetic packages, it became methods of assessing their reliability from solely thermally accelerated ageing towards the develpopment of methods to test for likely hazards due to the non-hermetic and possibly contamination-ridden encapsulating medium that was placed in direct contact with the semiconductor chip. Permeation of moisture and associated leaching of ionic contaminants from the plastic are hazards likely to arise from plastic encapsulation. This has been confirmed by reliability assessments involving the exposure of plastic encapsulated components to moist environments in humidity chambers, presure cookers and humid climates (references 2 and 3).

Semiconductor component lifetimes vary inversely with the exponential of RH (Figure 6), which appears to apply over a range of humidities from less than 30% to 90% RH. The rate or reaction (R_H) is expressed as follows:

(R_H) is expressed as follows:

$$R_H = const \ x \ exp(X.RH^2) \tag{4}$$

The expressions for acceleration by humidity (equation 4) and temperature (equation 3) may be combined to provide the following generalised expression for acceleration bydamp heat stress:

$$A = exp\left[X(RH_s^n - RH_{amb}^n) + Y(1/T_{amb} - 1/T_s)\right] \tag{5}$$

in which X and Y are constants and the suffixes 's' and amb' refer to stress and ambient respectively, and 'n' is usually 2. Clearly, increased humidities and temperatures provide higher acceleration factors. However, for the expression to remain valid and for authentic acceleration of ageing, it is essential that the vapour does not change state, i.e. the RH should be below 100% and definitely not saturated.

Acceleration Factors for Thermal and Humidity Stress

Acceleration factors for thermal overstress may be calculated by substituting a known value of activation energy that is relevant to the failure mechanism(s) in question and anchoring Tamb at the known or assumed normal operating (ambient)

temperature. Hence the acceleration achieved by stress at a particular elevated
temperature may be calculated. As optimistically high acceleration factors can
be obtained by anchoring Tamb at the lower operating temperatures, it has been
normal practice to assume the worst case operating temperature, thereby ensuring
that the thermal overstress test embraces even permanent worst-case operation.
Acceleration by purely thermal overstress has been found to be adequately covered by a value of $Y = Ea/k$ of 10.44×10^3, which gives the following acceleration
factors (Table 1) and test durations to represent 20 years life at a permitted
worst case ambient temperature of 70 C.

Table 1

Acceleration Factors and Test Durations for Thermal Overstress
Tests to Simulate 20 years Operation at 70 C

Test Temperature ($^\circ$C)	Acceleration Factor	Test Duration (Hours)
125	65	2700
150	315	600
160	555	320

Whilst Table 1 gives a set of acceleration factors and test durations associated
with a particular activation energy and operating environment, component failure
mechanisms can correspond to a wide range of failure mechanisms and acceleration
factors, and test conditions may have to be calculated for other operating
temperatures.

Acceleration factors due to damp heat overstress may be calculated by
substituting appropriate values for X and Y in equation 5, and anchoring the
expression at the operating ambient condition(s). Values for X have been found
to be spread about an average of about 5×10^{-4} (for RH given in per cent), and
values of Y ranging from about 7000 up to about 1,1000, dependent on the plastic
material and the humidity, have been measured for failure mechanisms occurring
in humid environments. Once again it is normal practice to employ the more
pessimistic values in order to embrace most conceivable failure mechanisms.
Consequently, values for X of 4.4×10^{-4} and Y of 7000 have been employed fo
calculate acceleration factors for damp heat stress applied typically at 85 C,
85% RH (humidity chamber) or 108 C, 90% RH (autoclave). Given in Table 2 are
the corresponding acceleration factors referred to a number of possible
telecommunications environments.

Thus, in most instances, both thermal overstress and damp heat stress do achieve
the considerable time-compressions (acceleration factors) sought from
accelerated ageing tests and such tests have been embodied in procurement
specifications required to meet 20 year lifetimes. The striking advantage of
autoclave testing, particularly to simulate uncontrolled tropical encivonments,
is very obvious. The significant benefits from reliability assessment by
relatively short duration tests have been the insights gained into causes of
unreliability, and the consequent improvement that has a been effected in
plastic encapsulating materials. Many manufacturers are now able to meet the
more stringent reliability specifications for plastic encapsulated semiconductor
components, and some even standardise on the reliability specifications as part
of routine testing to prove the reliability of their components.

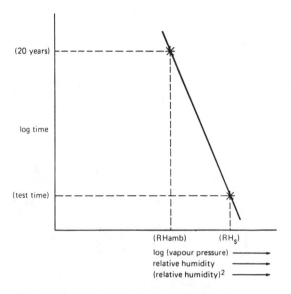

Moisture induced acceleration of ageing and its use to select overstress test conditions to simulate 20 years operation at ambient humidities.

Figure 6

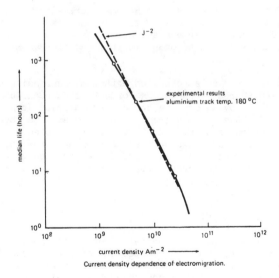

Current density dependence of electromigration.

Median life vs current density showing J^{-2} dependence of electromigration.

Figure 7

The expressions quoted apply to integrated circuits whether they are independently packaged or integrated within a hybrid microcircuit, and the factor that dominates will depend very much on whether it is humidity or temperature that prevails in the operating environment.

Table 2

Acceleration Factors and Test Durations for Damp Heat Overstress
Tests to Simulate 20 Years Operation in Various Environments

Environment		Overstress Conditions			
(Averaged over 10 years)		108 C, 90% RH		85 C, 85% RH	
		Acceleration Factor	Time (h)	Acceleration Factor	Time (h)
UK Telephone Exchange	30 C,25%RH	3100	60	600	300
UK Office	20 C,45%RH	3700	50	720	250
UK Uncontrolled	12 C,72%RH	1800	100	350	500
Tropic Uncontrolled	35 C,90%RH	90	2000	18	10,000

Accelerated Ageing by Electrical Overstress

Electrical overstress may be increased either by increasing current or voltage in the circuit, and both have been examined.

Current-induced failure mechanisms have become increasingly evident as ICs have grown in complexity and diminished in size, because the finer geometries have had to carry increasingly higher current densities which have caused the phenomenon of electromigration to occur in the conductors. Electromigration was covered in the previous chapter.

The failure mechanisms can be accelerated either by increasing temperature while maintaining a chosen current through the conductors (conventional Arrhenius plot, cf. Figure 2), or by increasing the current density at a fixed temperature (Figure 7), or both. However, the theory can be invalidated by excessive dissipation which generates significant temperature gradients in the conductors, or by carrying out accelerated ageing tests at temperatures above about 180 C (for aluminium). Therefore, the validation limits for accelerated ageing of electromigration are a current density of about 10^{10} A.m^{-2} and a temperature of about 180 C.

Accelerated Ageing of Thick Films

Thick Film circuit resistors are essential elements of hybrid microelectronics. Therefore it was necessary to establish methods of accelerating the normal ageing of these elements. Extensive studies of the reliability of thick-film resistors (references 4 and 5) have confirmed that their ageing behaviour also obeys the Arrhenius relationship, but with an activation energy different from that for semiconductor components. Acceleration by humidity has also been confirmed, but with an RH index of 1 and not 2. Hence, n = 1, X = 0.025, and Y = 8120, giving a combined expression for acceleration by damp heat as follows:

$$A = \exp\left[0.025(RHs - RHamb) + 8120(1/Tamb - 1/kTs)\right] \qquad (6)$$

in which RH is expressed as a percentage.

Ageing Behaviour of Thick-Film Resistors (TFRs)

The most common ageing behaviour of TFRs is a progressive increase in resistance value that varies approximately with the square-root of time. The changes increase with temperature and humidity. Most commercially obtained thick-film resistors drift by less than 0.5% in 20 years, and some undergo less than 0.1% drift.

SCREENING FOR RELIABILITY (RELIABILITY INDICATORS)

As discussed earlier, one method of assuring the reliability of components at the outset is to determine the criteria for acceptance or rejection of critical properties that are known to contribute to their unreliability. The specification of such tolerable limits has necessitated the study of the failure mechanisms, including the effects of the materials and operation of the components.

Hermeticity and Moisture Ingress

Hermeticity is a parameter that received attention from the earliest days of microelectronics packaging. Hermeticity is specified in terms of acceptable leak rates. More recent attention has been devoted to the ingress of water vapour, which has been found to be a serious hazard to semiconductor components - causing corrosion and parametric changes, especially in the presence of ionic contaminants, as described earlier. Based on the corrosion of CMOS ICs, the maximum quantity of water vapour that is tolerable in a hermetic package is about 5000 ppm which, based on the ambient conditions to be encountered, corresponds to a maximum equivalent helium leak rate of less than 10^{-10} atm.cm^{3}s^{-1} for a package volume of 0.1 cm^3. Proportionally larger leaks are tolerable for the larger hybrid packages.

At the higher external vapour pressures, the time to reach the critical level is only one day for a leak of 10^{-7} atm.cm^{3}s^{-1} and about 6 days for a leak rate of 10^{-8} atm.cm^{3}s^{-1}.

Reliability Indication by Electrical Measurement

Employing electrical parameters for reliability screening requires evidence of correlation with later failures and a good understanding of the mechanisms of failure.

Marginal Voltage Analysis (MVA)

MVA has been researched as a method for reliability screening of digital circuits (reference 6). Marginal voltage measurements can be used for either combinational or sequential circuits. Only the technique used for screening combinational circuits will be mentioned here. The basic set-up is very simple. The digital circuit under test is driven by a pattern generator that can be

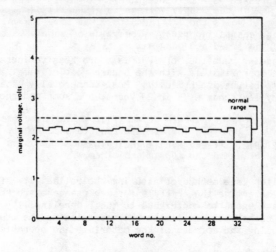

Typical marginal voltages for a TTL 2-bit adder

FIGURE 8

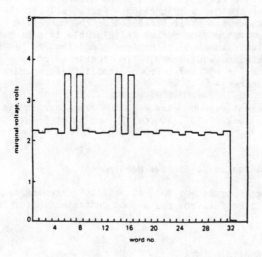

Anomalous marginal voltage pattern for TTL
2-bit adder

FIGURE 9

144

programmed to input any chosen word to the circuit. If, for example, the combinational circuit has 5 inputs there will be 32 input binary words associated with this circuit. For each input word the output is monitored for correct response whilst the supply voltage is gradually decreased from its nominal value (e.g. 5 volts for TTL circuits). At some low value of the supply voltage the output becomes incorrect. The value of supply voltage where this happens is termed the immediate marginal voltage.

Typically a pattern of marginal voltages results such as that shown in Figure 8, which shows the normal range of marginal voltages for this particular circuit type, a TTL 2-bit adder.

In some cases, however, the pattern of marginal voltages becomes anomalous or suspect as shown in Figure 9. The circuit functions correctly when the nominal supply voltage is used. It has been shown that such anomalies in nearly every case are caused by flaws in the integrated circuit.

The circuit may also be scanned by a light spot or electron beam synchronised with the electrical exercising in order to locate the fault site. Thus MVA is not only a method for indicating potential unreliability, but also a useful method for failure analysis.

REFERENCES
1.   Sinnadurai F N, 'Mechanisms and Modes of Failure in Licon Planar Semiconductor Devices', Published by British Post Office, March (1970)
2.   Sinnadurai F N, 'The Accelerated Ageing of Plastic Encapsulated Semiconductor Devices in Environments Containing a High Vapour Pressure of Water', Microelectronics Reliability, Vol 13, No 1, p 23, February (1974)
3.   Sinnadurai F N, 'Handbook of Microelectronics Packaging and Interconnection Technologies', Electrochemical Publications, Barns Street, Ayr, Scotland, UK, (1985)
4.   Sinnadurai N, Spencer P E and Wilson K J, 'Some Observations on the Accelerated Ageing of Thick-Film Resistors', Proceedings of the ISHM European Hybrid Microelectronics Conference, p 113 (1979)
5.   Sinnadurai N and Wilson K J, 'The Ageing Behaviour of Commercial Thick-Film Resistors', IEEE Transactions Components, Hybrids and Manufacturing Technology CHmt-5, p 308 (1982)
6.   Ager D J, Henderson J C, 'The Use of Marginal Voltage Measurements to Detect and Locate Defects in Digital Microcircuits', IEEE Proc IRPS, pp 139 -148, (1981)

10

Process Plant and Power Systems Reliability

R. N. ALLAN
**Department of Electrical Engineering
and Electronics
UMIST
Manchester, UK**

INTRODUCTION

This chapter describes several aspects concerning power-system reliability, including both generation and distribution. The concepts behind the techniques are applicable for the systems operated by utilities and those which exist within major industrial process plant complexes, e.g. oil refineries, chemical process plants. Sections of this Chapter have been reproduced with permission from Reference 2.

A power system serves one function only and that is to supply its customers or load points with electrical energy as economically as possible and with an acceptable degree of reliability and quality.

Design, planning and operating criteria and techniques have been developed over many decades in an attempt to resolve and satisfy the dilemma between the economic and reliability constraints. The criteria and techniques first used in practical applications, however, were all deterministically based.

The essential weakness of deterministic criteria is that they do not respond to nor reflect the probabilistic or stochastic nature of system behaviour, of customer demands or of component failures.

The need for probabilistic evaluation techniques has been recognised since at least the 1930's. These have been continually developed and today most utilities have reliability databases, computing facilities are greatly enhanced, reliability evaluation techniques are highly developed [1-5] and most engineers have a working understanding of probabilistic techniques. Consequently, there is now no need to artificially constrain the inherent probabilistic or stochastic nature of a power system into a deterministic one.

POWER SYSTEM RELIABILITY

The concept of power-system reliability is extremely broad and covers all aspects of the ability of the system to satisfy the consumer requirements. A simple but reasonable subdivision of

146

system reliability is system adequacy and system security. These two terms can best be described as follows.

Adequacy relates to the existence of sufficient facilities within the system to satisfy the consumer load demand. These include the facilities necessary to generate sufficient energy and the associated transmission and distribution facilities required to transport the energy to the actual consumer load points. Adequacy is therefore associated with static conditions which do not include system disturbances.

Security relates to the ability of the system to respond to disturbances arising within that system. Security is therefore associated with the response of the system to whatever perturbations it is subject. These include the conditions associated with both local and widespread disturbances and the loss of major generation and transmission facilities.

It is important to realise that most of the probabilistic techniques presently available for power-system reliability evaluation are in the domain of adequacy assessment

FUNCTIONAL ZONES AND HIERARCHICAL LEVELS

The basic techniques for adequacy assessment can be categorised in terms of their application to segments of a complete power system. These segments are defined as functional zones: generation, transmission and distribution. Adequacy studies can be, and are, conducted individually in these three functional zones.

The functional zones can be combined [6] to give the hierarchical levels shown in Figure 1. These hierarchical levels can also be used in adequacy assessment. Hierarchical level I(HLI) is concerned only with the generation facilities. Hierarchical level II (HLII) includes both generation and transmission facilities and HLIII includes all three functional zones in an assessment of consumer load point adequacy. HLIII studies are not usually done

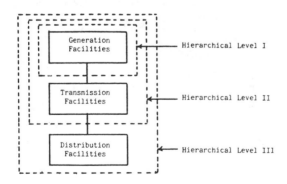

Figure 1   Hierarchical levels

directly due to the enormity of the problem in a practical
system. Instead, the analysis is usually performed only in the
distribution functional zone, in which the input points may or
may not be considered fully reliable.

GENERATING CAPACITY RELIABILITY EVALUATION

Concepts

At HLI the total system generation is examined to determine its
adequacy to meet the total system load requirement. This is
usually termed 'generating capacity reliability evaluation'.

In HLI studies, the reliability of the transmission and its
ability to move the generated energy to the consumer load points
is ignored. The only concern is in estimating the necessary
generating capacity to satisfy the system demand and to have
sufficient capacity to perform corrective and preventive
maintenance on the generating facilities. Deterministic criteria
have now been largely replaced by probabilistic methods which
respond to and reflect the actual factors that influence the
reliability of the system.

Criteria such as loss of load expectation (LOLE), loss of energy
expectation (LOEE), and frequency and duration (F&D) are now
widely used by electric power utilities. These indices are
generally calculated using direct analytical techniques although
sometimes Monte Carlo simulation is used. For full details of
evaluation techniques see Reference 2.

Loss of Load Expectation Method (LOLE)

The LOLE approach is by far the most popular and can be used for
both single systems and interconected systems [2].

The basic modelling approach for an HLI study is shown in Figure
2. The capacity model can take a number of forms. It is generally
formed in the direct analytical methods by creating a capacity
outage probability table [2]. This table represents the capacity
outage states of the generating system together with the
probability of each state. The load model (Figure 3) can either
be the daily peak load variation curve (DPLVC), which only
includes the peak loads of each day, or the load duration curve
(LDC) which represents the hourly variation of the load.

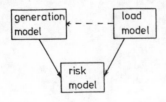

Figure 2   Conceptual tasks

148

Generally the DPLVC is used to evaluate LOLE indices giving a risk expressed in number of days during the period of study when the load will exceed available capacity. If an LDC is used, the units will be the number of hours. If LOEE indices are required, the LDC must be used.

It is clear from Figure 3 that a capacity outage of less than the reserve will not cause a loss of load. If O_K is the k-th outage state in the system capacity probability table, p_K the probability of this k-th outage and t_K is the number of time units for which this outage causes loss of load, then the contribution to the system LOLE by outage O_K is $p_K t_K$ time units. With n capacity outage states, the total LOLE, i.e. system risk, for the time interval being studied is [2]:

$$LOLE = \sum^n p_K t_K \quad \text{time units}$$

If the load characteristic is the DPLVC for a year, the units of LOLE are days/year. If it is a LDC for a year, the units of LOLE are hr/yr.

Because it takes a considerable period of time for a generating system to be planned, constructed and commissioned, it is essential to predict the need for it by using expansion planning studies. The loss of load expectation method can be an extremely valuable component of such studies.

To do this, the expected load growth must be predicted, the accepted system risk level must be determined and the size and type of units that can be added must be decided upon in advance.

CASE STUDY OF GENERATION EXPANSION

Systems Studied

The HLI techniques have been applied [7,8] to a case study based on three utilities serving a region in Saudi Arabia each of which needs to be reinforced to meet future predicted loads. The benefits of reinforcing separately or reinforcing by interconnecting the systems were studied. The basic data used for these systems are given in References 7 and 8.

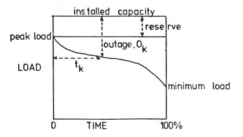

Figure 3   Load models

The studies centred on expanding the three utilities (A,B,C) over a 30-year period both as independent systems and as a single integrated system. Studies of this type enable the benefits, if any, that may accrue from integrated rather than dispersed systems, to be explored as well as deciding viable generation expansion plans. The results presented are limited to one or two scenarios only. However, in practice, many such scenarios would be studied, including variations in load growth, costs, FOR, unit sizes, acceptable risk level, installation dates, etc.

In these studies, it was assumed that 600MW units having an FOR of 0.01 would be added whenever the LOLE exceeded a specified risk of 0.11 day/yr.

Model used for Expansion Planning

The capacity planning model used [7,8] has the following main features:
(a) The existing system is taken as the starting point for the expansion analysis. It uses relevant number of units, types, capacities and forced outage rates (FOR).
(b) The capacity model and LDC for each year of the study are convolved to give [2] the LOLE, expected demand not served (EDNS), expected energy not supplied (EENS) and expected index of reliability (EIR).
(c) The evaluated LOLE is compared with a specified acceptable risk level (R). If the LOLE is greater than R, one unit is added at a time until the LOLE is less than or equal to R. This creates the expansion plan and the capacity model is updated for each addition.
(d) The present value (PV) cost of any units added including system capital costs and system operation and maintenance costs and the PV of the outage costs associated with both the EDNS and EENS are evaluated for each year of the planning study.

Certainty and Uncertainty in Load Forecasting

A 30-year expansion plan for systems A, B and C, assuming certainty and uncertainty in forecasting the demand, were determined. It was assumed that the uncertainty could be represented by a normal distribution having a standard deviation of 15% and that this remained constant for the planning period. The risk incorporating uncertainty was evaluated using the conditional probability approach [1,2]. A summary of these expansion plans are shown in Table 1.

It can be seen that, for both the certainty and uncertainty cases, the number of units and total PV cost is reduced if the three systems are consolidated as one, although the benefit is less in the uncertainty case; the reduction in the total PV cost is 25% and 14.6%, respectively. This is not the overall saving, however, because the system must be linked together in order to create an integrated system. The next stage must, therefore, assess whether the above savings justify the expenditure on the required network.

The second observation that can be made from Table 1 is that,

with load forecast uncertainty, an increased number of units, 31
against 23, are needed. This causes an increase of 75% in total
PV system cost. Therefore, peak demand uncertainty is an
extremely important parameter and there is clearly a crucial link
between demand forecasts and capacity decisions in order to meet
the load with an acceptable level of risk.

Table 1   Effect of Uncertainty in Load Forecasting

| System | with certainty | | with uncertainty | |
	no of units	PV cost £M	no of units	PV cost £M
A	13	551.6	17	721.2
B	10	441.1	12	529.5
C	3	160.8	4	196.6
Total	26	1153.5	33	1447.3
Combined	23	919.9	31	1263.3

Uncertainty in Unit Installation Time

The possibility of delaying or advancing the unit installation
date should be considered in planning, either because of it
having a worthwhile economic benefit or because delays can occur
unexpectedly in practice and its effect should be studied. A
summary of the results are shown in Table 2, which indicates the
effect on the individual system expansions and the integrated
system expansion plan caused by delaying or advancing the
installation dates by one year.

Table 2   Effect of Installation Dates

| syst | certain date | | delayed date | | advanced date | |
	SC	OC	SC	OC	SC	OC
A	551.6	2.79	504.9	14.78	577.8	1.29
B	441.1	2.84	421.1	9.94	462.1	1.37
C	160.8	1.97	153.5	2.68	168.5	1.73
tot	1153.5	7.60	1079.5	27.40	1208.4	4.39
comb	919.9	4.37	834.8	89.57	963.7	1.38

where SC = system cost (£M)
       OC = outage cost (£M)

It is evident from Table 2 that, if the installation date of a
unit which should be installed in a specific future year is
delayed until the next year, the risk level increases.
Consequently, the PV system cost decreases because of payment

151

postponement but the PV outage cost increases due to the deterioration of risk. In the case of the combined system, a delay of one year decreases the system cost by 9.3% and increases the outage cost by 1950%. On the other hand, an advance of one year increases the system cost by 4.8% and decreases the outage cost by 68%

Sensitivity Analyses

Sensitivity analyses form a vital part of the practical application of planning in order to understand and appreciate the impact that various uncertainties have on the economics and reliability of the system. The following results illustrate some sensitivity studies from an analysis of system B only.

The effect of delays in excess of one year was studied. The effect on risk index is shown in Figure 4. It is seen that delays

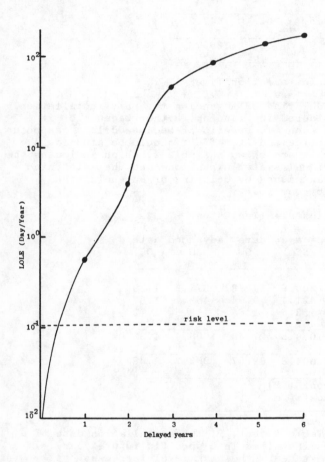

Figure 4   Effect of delay on LOLE

152

have a considerable effect on reliability. This increase in risk explains the rise in outage costs resulting from delays in unit installation. Although the outage costs increase rapidly as the delay is increased, the system cost steadily decreases.

LOLE is dependent on the generating unit availability parameters and, in practice, these are based on available data and future forecasts. The uncertainty in these parameters, therefore, affects reliability and costs. This impact is shown in Figure 5, which indicates the number of units required, the associated outage costs and the effective unit capacity as a function of FOR. These results show that the FOR can produce very significant effects on both reliability and cost.

The effect of adding different sizes of units was studied. The variation of LOLE with each unit size is shown in Figure 6. The system risk varies smoothly with the 100MW units, since these are

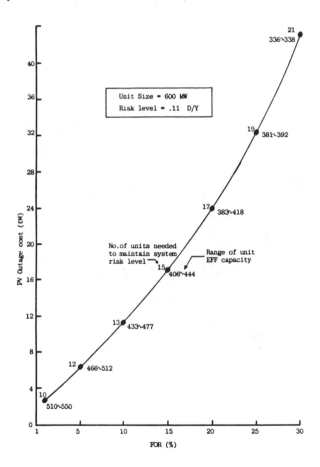

Figure 5   Effect of forced outage rate

added regularly to the system each year. The variation of risk for the other units is very similar because capacity additions occur at similar time. In long-range planning, large units often appear to be economically advantageous since they are generally cheaper to operate. Economic evaluation of alternative sizes should, however, include the effect on system reliability. In practice, mixing of generation, both size and technology, would also be studied before a final expansion plan was approved.

Summary Comments of Case Study

Although the results of this case study reveal that major benefits can accrue by consolidation of the three systems in terms of the required generating capacity, the savings will be offset against the capital and running costs of the interconnections required to link the three systems. This aspect would form the next stage of the planning process. In order to evaluate the economic worth, the performances of the whole system, with and without interconnection, must be assessed. This would determine the trade-off between the saving in generation cost and the cost of the interconnections.

COMPOSITE SYSTEM RELIABILITY EVALUATION

Concepts

In HLII studies, the simple generation-load model is extended to include bulk transmission. Adequacy analysis at this level is usually termed composite system or bulk transmission system evaluation.

HLII studies are required to assess the adequacy of an existing system and the impact of various reinforcement schemes, at both

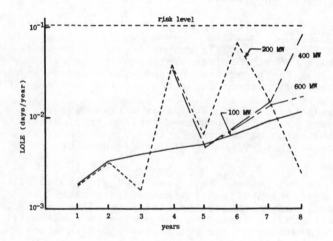

Figure 6   Effect of unit size

154

the generation and transmission levels. These effects can be assessed by evaluating two sets of indices; individual bus (load-point) indices and overall system indices. The system indices give an assessment of overall adequacy and the load-point indices monitor the effect on individual busbars and provide input values to the next hierarchical level. The basic evaluation techniques for HLII evaluation are well documented [2].

There are however many complications in this type of analysis such as overload effects, redispatch of generation and the consideration of independent, dependent, common-cause and station-associated outages. Many of these aspects have not yet been fully resolved and there is no universally accepted method of analysis. This section considers one particular case study associated with terminal effects.

Terminal Station Effects

The main problem area in HLII studies is the enormous amount of computational effort required if an exhaustive evaluation is made of the complete system representation.

In order to simplify the problem, the circuits are usually represented as single elements and the terminal stations (substations and switching stations) as single busbars. The terminal stations are therefore not fully represented and the switching and breaker configurations are neglected. The detailed analysis then proceeds by considering combinations of line and generator outages as independent overlapping events. This approach is reasonable provided no events can occur which cause the simultaneous outage of more than one line or generator.

Failure events inside the station can however cause one or more lines/generators to be outaged simultaneously. These are defined [9] as station originated outages.

The types of failure events which can occur in a terminal station are:
a) active failures. These are usually ground faults on breakers or busbars which cause the appropriate protection breakers to operate. Depending upon the protection scheme, the breakers which trip can simultaneously disconnect more than one line and/or generator.
b) stuck breaker conditions. If, following an active failure, one or more of the primary protection breakers fail to trip, other breakers will respond which may disrupt a larger section of the station and may cause a greater number of lines/generators to be disconnected.

These two concepts are not new and have been used widely [10,11] in the reliability evaluation of substations and distribution systems. Correct modelling of the system, the ways in which it operates and can fail and the impact on the system is fundamental in the reliability evaluation of the composite system in which the terminal station exists.

One assumption often made [9,10,11] is that the probability of more than one simultaneously stuck breaker is negligible.

155

Although this may be true in many instances, it is an assumption that cannot be justified without detailed examination. It will be a reasonable assumption if each breaker is actuated quite independently of all others. On the other hand, protection schemes sometimes have common components, e.g. the fault detection device, battery supplies, etc. If any of these common components fail, the entire protection scheme will fail and more than one breaker may be "stuck" simultaneously.

The following sections consider a case study which illustrates typical failure events that can occur in a terminal station, describes how the indices of these events can be assessed by performing a reliability analysis on the protection system of the terminal station and demonstrates how these events and their associated indices can be included in HLII studies.

CASE STUDY ON TERMINAL EFFECTS

Terminal Station Outage Effects

The single line diagram shown in Figure 7 is a ring switching station.

A detailed failure modes and effects analysis of the system is first required. A selected number of events for the system is shown in Table 3 in which the symbol A represents an actively failed component, S represents a stuck breaker and M represents a maintenance outage which may be caused either by a scheduled maintenance or by the component requiring repair following a previous failure event. A more detailed list is given in Reference 9.

Table 3   Selected Station Outages

failure mode	lines/generators outaged
1. 2A	G3, L1
2. 1A + 2S	G3, G4, L1
3. 2A + 1S	G3, G4, L1
4. 2A + 3S	G2, G3, L1
5. 2A + 1S + 3S	G2, G3, G4, L1
6. 1M + 2A	G3, L1
7. 3M + 1A	G3, G4, L1
8. 3M + 2A	G3, L1
9. 4M + 2A	G2, G3, L1
10. 5M + 2A	G1, G2, G3, L1
11. 6M + 2A	G3, G4, L1

The general practice used in composite system reliability evaluation is to consider the contingencies such as shown in column 2 of Table 3 as made up of component outages which are independent.

This is not a realistic approach when the indices of the overlapping events or combinations are evaluated assuming independent component outages because the method does not recognise the fact that a particular station outage may remove the selected combination of components by a single station event.

As an example of this problem, consider the information provided in Table 3. This shows that the active failure of breaker 2 removes G3 and L1 simultaneously. The indices of this event should therefore include not only those associated with the independent overlapping outage of G3 and L1 but also those associated with the event that initiates the simultaneus outage of G3 and L1, i.e. the indices associated with the active failure of breaker 2.

Inclusion of Terminal Station Events

A detailed analysis that provides the information shown in Table 3 can not be accommodated within a single composite system evaluation because of the large computational effort. Instead, a separate assessment of the station can be performed before commencing the composite system reliability evaluation. This will produce a list of contingencies that should be studied, a ranking order in terms of severity or probability of occurrence and the relevant indices of the event.

Table 3 shows that several failure modes lead to the same system effect, i.e. they cause the same set of lines and generators to be outaged. For example, failure modes 2,3,7 and 11 all lead to the simultaneous outage of G3, G4 and L1. It follows therefore that these groups and their indices can be combined together for the purposes of composite reliability evaluation.

The remaining problem is therefore to evaluate the indices of each of the contingencies to be assessed. These indices will depend upon whether the protection system operates as desired or whether one or more of the protection breakers fail to operate, i.e. are stuck. For instance, the outage of G3 and L1 (event 1 in Table 3) will occur with a rate equal to the failure rate of breaker 2 weighted by the probability of both breakers 1 and 3 operating successfully, the outage of G3, G4 and L1 (event 3 in Table 1) will occur with a rate equal to the failure rate of breaker 2 weighted by the probability of breaker 1 failing to operate and breaker 3 operating successfully.

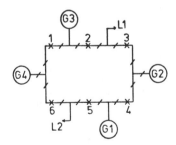

Figure 7   Ring switching station

It follows that the indices of each outage can be deduced from a knowledge of the reliability indices of each station component and a knowledge of the probabilities of the protection system operating and not operating. The last aspect was the subject of the Chapter on System Reliability Modelling concerning event trees. This previous information can be used directly in the present case study.

HLII Analysis Including Terminal Events

The contingencies to be included should include those caused by terminal station outages in addition to the conventional overlapping component outages. The indices therefore consist of two distinct contributions. These are:
a) the direct contribution associated with the overlapping events causing the contingency
b) the indirect contribution associated with terminal station outages.

The second contribution is deduced from the component reliability indices of the initiating event (station component failure) and the probability of the protection system operating or not operating. As an example, consider Table 3, the protection probabilities shown in the Chapter on System Reliability Modelling and the following contingencies caused by the active failure of breaker 2. Let this breaker have an active failure rate of λ^A of 0.04 f/yr.

a) contingency G3, L1 (event 1):
λ(G3,L1) = λ^A(breaker 2) x P(breakers 1 and 3 both operate)
$\qquad$ = 0.04 x 0.984754 = 3.94×10^{-2} f/yr

b) contingency G3, G4, L1 (event 3):
λ (G3,G4,L1) = λ^A(breaker 2) x P(breaker 1 fails and breaker 3 operates)
$\qquad\qquad$ = 0.04 x 0.001974 = 7.90×10^{-5} f/yr

c) contingency G2, G3, G4, L1 (event 5):
λ(G2,G3,G4,L1) = λ^A(breaker 2) x P(breakers 1 and 3 fail)
$\qquad\qquad$ = 0.04 x 0.011298 = 4.52×10^{-4} f/yr

This concept can be applied to all the contingencies in Table 3. Each contingency is then simulated in a composite reliability assessment using standard techniques [2].

Summary Comments on Case Study

This case study has shown the importance of realistically correct modelling procedures. It has shown that the contingencies which should be studied can be made significantly more severe by protection system operation and failures in the terminal station; thus having a greater impact on the overall behaviour of the composite system.

It has also demonstrated the practical use of the event tree method in the reliability evaluation of protection systems.

DISTRIBUTION SYSTEM RELIABILITY EVALUATION

Concepts

The overall problem of HLIII evaluation can become very complex
in most systems because this level involves all three functional
zones. Therefore, the distributional zone is usually analysed as
a separate entity. The HLIII indices can be evaluated, however,
by using the HLII load-point indices as the input values at the
sources of the distributional zone being analysed.

The models and techniques described in the Chapter on System
Reliability Modelling allow the three basic reliability indices,
expected failure rate (λ), average outage duration (r) and
average annual outage time (U), to be evaluated for each load
point of any meshed or parallel system. These techniques are also
applicable to most network (flow) problems including
communications, water, gas, etc. They have three major
deficiencies, however:
(a) they cannot differentiate between the interruption of large
and small loads;
(b) they do not recognise the effects of load growth by existing
customers or additional new loads;
(c) they cannot be used to compare the cost-benefit ratios of
alternative reinforcement schemes nor to indicate the most
suitable timing of such reinforcements.

These deficiencies can be overcome by the evaluation of two
additional indices, these being:
(i) the average load disconnected due to a system failure,
measured in kW or MW and symbolized by L;
(ii) the average energy not supplied due to a system failure,
measured in kWh or MWh and symbolised by E.

These are not new indices and techniques for their evaluation are
well documented [2].

The basic criterion for determining a load point failure event
was 'loss of continuity', i.e. a load point fails only when all
paths between the load point and all sources are disconnected.
This assumes that the system is fully redundant and any branch is
capable of carrying all the load demanded of it. This clearly is
unrealistic. For this reason, the previous 'loss of continuity'
criterion is best described as 'total loss of continuity' (TLOC).
In addition, a system outage or failure event many not lead to
TLOC but may cause violation of a network constraint, e.g.
overload or voltage violation, which necessitates that the load
of some or all of the load points be reduced. This type of event
is defined as partial loss of continuity (PLOC). The evaluation
of PLOC events becomes of great significance if the load and
energy indices are to be evaluated. The relevant detailed
techniques needed to evaluate these events and the load and
energy indices are described in Reference [2,12].

Total Loss of Continuity (TOLC)

The equations for λ, r and U of TLOC are given in the previous Chapter. The values of L and E are readily evaluated knowing only the average load connected to each load point since

$$L = L_A$$
$$ = L_P f$$
$$E = LU$$

where

L_A = average load at load point
L_P = peak load at load point (maximum demand)
 f = load factor
and U = annual outage time of load point.

Partial Loss of Continuity (PLOC)

A partial loss of continuity event could potentially occur for any combination of branch and busbar outages except those that cause a total loss of continuity. In order to be rigorous, it would therefore be necessary to simulate all possible outage combinations except those that are known to lead to a TLOC event. This may be feasible for very small systems but it becomes impractical for large ones.

It is usually feasible to study all first order outages and usually reasonable to neglect third and higher order outages.

After selecting the outage combinations to be considered, it is necessary to deduce whether any or all of these form a PLOC event. This can only be achieved using a load flow and establishing whether a network constraint has been violated.

The reliability indices of each load point of interest due to a PLOC event are given by [2,12]

$$\lambda = \lambda_E P + \lambda_E (1 - P)\lambda_L \ \frac{r_E r_L}{r_E + r_L}$$

$r = r_E$ if load remains disconnected until
       repair is complete

$r = \dfrac{r_E r_H}{r_E + r_H}$ if load is disconnected and
            reconnected at each load transition

$U = \lambda r$

$L = [\int_{o}^{t_1} L(t)dt - L_S t_1 \ ]/t_1$

$E = LU$

where
λ_E = rate of occurence of outage condition
μ_E = reciprocal of the average duration r_E of the outage condition
L_S = maximum load that can be supplied to the load point of interest during the outage condition
P = probability of load being greater than L_S

λ_H = transition rate from load levels > Ls to load levels ≤ Ls
= reciprocal of average duration (r_H) of load level > Ls
λ_L = transition rate from load levels ≤ Ls to load levels > Ls
= reciprocal of average duration (r_L) of load level ≤ Ls
L(t) = load-duration curve
t_1 = time for which the load level > Ls.

Numerical Example

In order to illustrate the effect of including a PLOC criterion, consider the system shown in Figure 8.

Using typical data [2], the results for TLOC and PLOC are shown in Table 4.

Table 4   Indices for system of Figure 8

load point	λ f/yr	r h	U h/yr	L MW	E MWh/yr
2					
TLOC	2.0×10^{-2}	5.0	1.0×10^{-1}	15	1.50
PLOC	3.6×10^{-6}	4.9	1.8×10^{-5}	4.6	8.1×10^{-5}
total	0.020	5	0.100	15	1.50
3					
TLOC	2.0×10^{-2}	5.0	1.0×10^{-1}	7.5	0.75
PLOC	7.1×10^{-2}	9.6	6.8×10^{-1}	3.3	2.24
total	0.091	8.6	0.784	3.8	2.99

It can be seen from Table 4 that the PLOC indices associated with load point 2 are very small, whereas those associated with load point 3 are very significant. It is evident therefore that, although PLOC may be insignificant for some load points of a system, they may dominate those of other load points. It follows that PLOC should be included in the reliability analysis of distribution systems in order to ensure accuracy of the

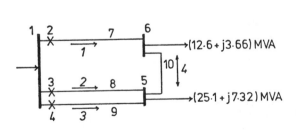

Figure 8   Ring distribution system

161

evaluation and the most reliable set of information necessary for the decision-making process of expansion and reinforcement.

CASE STUDY OF DISTRIBUTION RELIABILITY

Practical Distribution Systems

The 33/11kV system shown in Figure 9 is based on a real UK distribution system. It illustrates the complexity of likely features in a system and serves as a useful example to demonstrate modelling and evaluation techniques.

Both the 33kV network and the 11kV network are shown, the main features being:
a) each 11kV feeder has a number of load-points distributed along

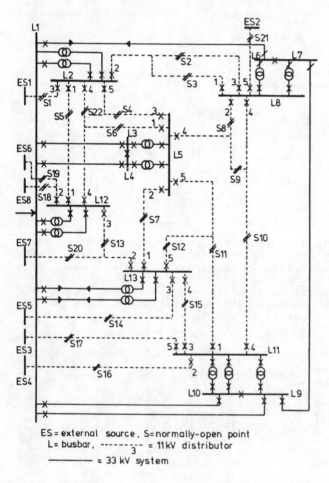

ES = external source, S = normally-open point
L = busbar, ------$\frac{}{3}$-- = 11kV distributor
———— = 33 kV system

Figure 9   Typical distribution system

its length
b) many of the 11kV feeders are connected through normally-open
points to other feeders fed from the same or other sources.
c) loads may be transferred from one substation to another by
disconnecting feeders at their normal source end and closing one
or more normally-open points. This is done only in the event of a
failure causing loss of supply at the source substation and it is
not normal practice to parallel sources
d) a failure such as in (c) may cause total loss of continuity in
which case all loads are disconnected until either repair is
completed or load is transferred or it may cause partial loss of
continuity because of unsatisfactory system conditions. In the
latter case, only sufficient loads necessary to regain
satisfactory system conditions are disconnected and these remain
disconnected until system conditions recover, repair is completed
or load is transferred.
e) several alternative switching actions may exist to transfer
the load of any given feeder.

From consideration of these features, it is evident that total
loss of continuity is an adequate criterion only for single
feeder radial networks capable of supplying the maximum demand.
Hence this criterion has limited application.

Analysis of the Typical System

To illustrate the TLOC and PLOC results that can be achieved from
the use of the previous models, the reliability indices of
load-point L5 of the system are shown in Table 5. The data used
for this system has been published elsewhere [12]. Two different
case studies were examined:
case A   - total loss of continuity
case B   - total and partial loss of continuity

Table 5   Reliability indices for load point 5

case	λ f/yr	r h	U h/yr	L MW	E MWh/yr
TLOC	0.00033	65.0	0.021	4.063	0.0875
PLOC	0.0028	322	0.900	0.602	0.5415
total	0.00313	294	0.921	0.682	0.6290

A considerable number of comments and conclusions can be drawn
from the results shown in Table 5 but only the most pertinent
will be discussed.

Considerable errors would have occurred in all of the indices if
partial loss of continuity events had been ignored. The latter
criterion considers many failure events which are not identified
by the total loss of continuity criterion. Therefore the failure
rate, the unavailability and the energy not supplied at the load
point of interest increases. The average load not supplied,

163

however, decreases because these additional events cause loads less than the average loading level of the busbar to be disconnected. Depending on the amount by which the failure rate and the unavailability increase, the average outage time may increase or decrease.

The load point indices of busbar 5 are dominated by the partial loss of continity criterion and considerable increases in λ, r, U and E are found. Therefore the effect of including the partial loss of continuity criterion is substantiated and its contribution to the total indices can be very significant.

CONCLUSIONS

This chapter has discussed various aspects concerning power system reliability. In particular, it has described some philosophical concepts and some adequacy evaluation techniques; the latter being supported by case studies. Both of these aspects will continue to be developed as more utlities apply the concepts to a wide range of system types in order to enhance objective decision making.

When applying these concepts and techniques, there should be some conformity between the reliability of various parts of the system. This does not mean that the reliability of each part should be identical and there are many reasons why they should be different. For instance, a particular load may be of great importance and needs high reliability. Also, failures in the generation and transmission systems can cause widespread outages and hence serious societal consequences whereas failures in the distribution system are usually very localised.

Finally, when improvement in reliability is being considered, there should be some benefit from such an improvement. The most useful method for assessing this benefit is to equate the incremental or marginal investment cost to the incremental or marginal consumers' valuation of the improved reliability. This is extremely difficult at the present time because of lack of data regarding consumers' valuation. However this data is being increasingly considered. In the meantime it is still very beneficial for individual utilities to decide and use a consistent criterion by which the benefit of various expansion and reinforcement schemes can be assessed.

REFERENCES

1. R.Billinton, R.N.Allan, "Reliability Evaluation of Engineering Systems - Concepts and Techniques". Longmans (formerly Pitmans), 1983.

2. R.Billinton, R.N.Allan, "Reliability Evaluation of Power Systems". Longmans (formerly Pitmans) 1984.

3. R.Billinton, "Bibliography on the Application of Probability Methods in Power System Reliability Evaluation". IEEE Trans, PAS-91, 1972, pp.649-660.

4. IEEE Committee Report, "Bibliography on the Application of Probability Methods in Power System Reliability Evaluation 1971-1977". IEEE Trans, PAS-97, 1978, pp.2235-2242.

5. R.N.Allan, R.Billinton, S.H.Lee, "Bibliography on the Application of Probability Methods in Power System Reliability Evaluation 1977-1982". IEEE Trans, PAS-103, 1984, PP.275-282.

6. R.Billinton, R.N.Allan, "Power System Reliability in Perspective". IEE Journal on Electronics and Power, 30, March 1984, pp.231-236.

7. A.M.Shaalan, "Reliability Evaluation in Long Range Generation Expansion Planning". PhD Thesis, UMIST, Manchester, 1984.

8. R.N.Allan, A.M.Shaalan, J.R.Ochoa Mendoza, "Cost Benefit and Reliability Assessment of Electrical Generating Systems". Reliability '85 Conference, Birmingham, 1985, paper 5C/3.

9. R.Billinton, T.K.P.Medicherla, "Station Originated Multiple Outages in the Reliability Analysis of a Composite Generation and Transmission System". IEEE Trans, PAS-100, 1981, pp.3869-3879.

10. M.S.Grover, R.Billinton, "A Computerized Approach to Substation and Switching Station Reliability Evaluation". IEEE Trans, PAS-93, 1974, pp.1488-1497.

11. R.N.Allan, R.Billinton, M.F.De Oliveira, "Reliability Evaluation of Electrical Systems with Switching Actions". Proc IEE, 123, 1976, pp.325-330.

12. R.N.Allan, E.N.Dialynas, I.R.Homer, "Modelling and Evaluating the Reliability of Distribution Systems". IEEE Trans, PAS-98, 1979, pp.2181-2189.

11

■

Reliability Concepts in Power System Planning

ABDULLAH SHAALAN
and NASSER AL-MOHAWES
Electrical Engineering Department
College of Engineering
King Saud University, Riyadh, Saudi Arabia

ABSTRACT

The problem of power system planning, due to its complexity and dimensionality is one of the most challenging problems facing industry in developing as well as developed countries. In the planning phase, system generating capacity is a crucial step in planning the expansion of modern electric power systems. In reliability-based generation expansion planning the most widely used reliability criterion is the loss-of-load expectation index (LOLE). However, to provide more information to judge the adequacy of power systems and to help in making proper decisions, other complementary indices have been developed. The method described above is based on a fast and approximate technique, which reduces the computational time for large systems without introducing significant error in the results.

INTRODUCTION

In our modern society, power system engineers are responsible for the planning, design and operation of complex power systems. Power system planning and reliability evaluation methods have been developed to aid engineers in these tasks (1 - 11). Power system failures can cause effects that range from inconvenience and irritation to a severe impact on society and the environment.

Evaluating generating capacity is becoming one of the most crucial steps in planning the expansion of modern electric power systems (4). Generating capacity reliability is defined in terms of the adequacy of the installed capacity to meet the system load level at all times. Hence, when outages occur (i.e. scheduled or unscheduled) a reserve must provide for this shortage in system capability to meet the peak load.

The basic methodology for system reliability evaluation is to develop probabilistic models i.e. capacity models and load models and calculate the probability of loss of load by a convolution process of the two models for all the intervals in the intended period (e.g. days in a year) taking into account the changes in load demand, scheduled outages, unit additions or deletions, size and type, pooling of systems, interconnected systems, and load forecast uncertainty. A summation of the period indices provides the annual index known as loss of load expectation (LOLE) in days/year. Other relevant reliability indices can be deduced with minor efforts.

DERIVATION OF RELIABILITY INDICES

Several indices have been used for reliability evaluation in power system planning. These include:

1. The loss of load expectation, LOLE (days/year).
2. The expected energy not served, EENS(MWH).
3. The energy index of reliability, EIR.

All of these indices are probabilistically based and yield differing, not strictly comparable, indices. The LOLE index is the most commonly used, especially in system planning. It yields measures of the form 'one day in 10 years' loss of load expectation, which is the criterion commonly adopted.

The expected energy not served, EENS, evaluates the expected energy curtailed due to deficiencies in system generating capacity.

The first step in deriving these indices is to define models that describe the situation of interest. There are two situations which need to be modelled, namely:

1. The available capacity (capacity model).
2. The load duration over a given period of time (load model).

The capacity model in its simplest form consists of representing each unit of capacity in the system by a 2-state model and characterising it as a random variable C_i. This has value 0 with a probability equal to its forced outage rate (FOR) and value C_i, with a probability equal to the complement of its FOR. Therefore, the capacity available to the system from unit 1 is

$$C_i = \begin{cases} 0, & P(C_i) = 0 = FOR = q_i \\ C_i, & P(C_i) = C_i = p_i \end{cases}$$

where $p_i + q_i = 1$

The point of interest is the total available capacity states of all units in the system:

$$C_t = \sum_{i=1}^{N} C_i$$

where C_t is another random variable and represents the total available system capacity. If the capacity states are assumed independent, then the probability distribution of C_t is obtained by convolving the probability distribution for each of the C_i, so the C_t density function can be developed as:

$$P(C) = \sum_{C_i=0}^{C_i=C} P(C-C_i) P(C_i)$$

$$= f * f_{C_i}$$

Loss of Load Expectation.(LOLE)

As previously mentioned, LOLE is the most widely used method in system expansion planning (2). This index is derived by combining the capacity model previously defined with an appropriate load model to obtain a measure of the reliability of the system. i.e. the adequacy of the amount of generation available in the system to meet the load and withstand expected but unscheduled future outages.

Assume the simplest case in which the load L on the system is constant and perfectly predictable. The system loss of load probability would be the probability of all outages that would cause the system effective capacity to be less than the system load L. If the system reserve margin is M, then:

$$M = C - L$$

The LOLP (the probability that system outage O_i is greater than system reserve margin M) is defined as:

$$LOLP = \sum_{O_i > M} P(O_i)$$

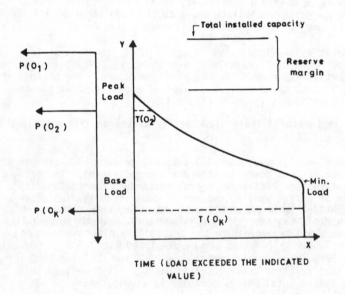

Figure 1: Relation between load, capacity and reserve.

In practical cases, however, the load is not constant during the period. The load model used in this method is the load characteristic shown in Figure 1. This load characteristic is a plot of the load level as a function of the number of time units which a given load level exceeds the indicated value. It may represent the hourly peak loads in a day (load duration curve), the daily peak loads in a year (daily peak load variation curve) or any other appropriate time interval of interest. The curve shows the peak load in the period of study and all the load levels organised in a descending order of magnitude. The lowest load level is referred to as the base load. The curve is interpreted as follows: for each load level on the vertical axis, the curve shows the number of time units during the period that the load will be exceeded.

The behaviour of M = C - L and the probability that M < 0 can now be investigated. If both C and L are independent random variables, the density of C - L is the convolution of the density of C with the density of (- L):

$$f_M = f_C * f_{-L}$$

This is the basic interpretation and formulation of the LOLP model. From Figure 1, it can be seen that any outage O_i (i = 1,....., K) greater than M will cause loss of load during a fraction of time $T(O_i)$ of the planning interval. If $P(O_i)$ is plotted along the vertical axis of system load capacity shown in Figure 1 with O_i, the complement of C increases as it approaches the horizontal axis. The value of O_k represents the event in which the total system capacity is out. Therefore, the total loss of load expected for the study interval is (3):

$$LOLE = \sum_{i=1}^{K} P(O_i)T(O_i) \text{ time units}$$

If the load characteristic is the load duration curve (LDC), the loss of load expectation is in hours. If a daily peak load varation curve (DPLVC) is used, the loss of load expectation is in days for the period of study. The period of study could be a week, a month or a year. The simplest application is the use of the curve on a yearly basis, as shown in Figure 1.

Expected Energy not Served, EENS

The EENS is becoming more widely used, and will possibly replace LOLE because of its physical significance in that power systems are energy systems. To derive a measure related to the expected fraction of energy not served due to capacity loss, the probabilities of having varying amounts of capacity unavailable are combined with the system load by means of the load duration curve. The significance of this index is that it can be related to revenue collected by a utility and purchases made by customers.

In order to derive the EENS, the capacity model developed for LOLE and EDNS is retained and a system load duration curve (LDC) is used. It can be seen that the area under the LDC is the total energy demanded by the system.

For any outage, O_i, greater than the reserve M, the amount of energy not served can be predicted from the LDC. (Figure 2)

From Figure 2, the energy not served, E_i, would be $O_i T_i$ (integrated area under the LDC). Therefore, the total expected energy not served, EENS, may be computed by the relationship (5):

169

$$EENS = \sum_i E_i P(O_i) \ MW$$

where E_i is the area under the LDC (Figure 2) representing the energy not served for a given outage of size O_i, and $P(O_i)$ is the probability of an outage of size O_i.

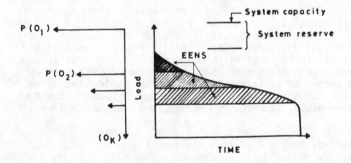

Figure 2: Shaded area under the LDC represents multiple energies not served due to multiple outages.

Energy Index of Reliability, EIR

In generation system planning, it is often useful to find the ratio between the energy not served and the total energy demanded by the system, because this provides a reliability index that does not depend on particular system characteristics. Hence, it can be compared with those of other systems with different generating capacity and different load demand. The EIR is an extended form of EENS discussed in the preceding section. To obtain EENS in per unit demanded ($EENS_{pu}$), EENS is divided by the total energy demanded (TED) during the given time period. The value of TED is represented by the total area under the load duration curve, i.e.

170

$$EENS_{p.u.} = \frac{EENS}{TED}$$

The $EENS_{p.u.}$ is usually very small because the energy curtailed is a small fraction of the total energy demanded. This is due to the fact that the energy lost is related to the outages and the time spent on outages, and both of these are small relative to the total levels of load and period of time. The results may also be expressed in terms of an energy index of reliability (EIR), by the relation (2,5):

$$EIR = 1 - EENS_{p.u.}$$

TECHNIQUES FOR EVALUATING RELIABILITY INDICES

The concept behind the LOLE and the subsequent measures of EENS and EIR discussed in the previous sections is based on theoretical techniques for constructing the capacity outage probability table, COPT. In theory, the generation capacity model contains all generating states together with their probabilities of occurrence. The basic technique for evaluating a COPT is given in basic reliability evaluation books (e.g. 2,3,4,6).

The conventional method for evaluating the COPT is to recursively add the units of the generating system one at a time until the capacity outage table includes all units of the system. This table is an exact table if all possible outages have been considered. In this case, the number of states is 2^N where N is the total number of units residing in the system. Usually, probabilities less than a certain value, say, 10^{-8}, are considered negligible for practical purposes and hence the corresponding states are neglected. This is known as truncation.

When a practical system contains a large number of units of different capacities, the exact method of deriving the COPT becomes tedious and time consuming and the table contains a large number of discrete capacity outage levels. Even using truncation, the required storage and computation time still remain excessive (7).

In order to reduce the size of the COPT to a reasonable size without introducing a significant error in the reliability indices, an approximate rounding technique is used. The method is briefly described as follows (7): The total capacity installed is divided so that a maximum number of states, N_M, exists in the COPT. This is achieved by choosing a step length (T) in MKW, whose integer multiples represent the capacity on outage. The new COPT will thus contain the outages: 0T, 1T, 2T,...$(N_M-1)T$, i.e. N_M states. The last state will be the first multiple of T greater than the total installed capacity of the system under consideration. Some of these outages have very small probabilities and can be ignored. This is achieved by truncating all values less than a certain accuracy limit.

In the rounding method, the unit capacity in the system will not generally coincide with the integer multiple of the step length T but will lie somewhere between two states. Consequently, the probability of having this capacity on outage is shared proportionately between the two rounded states immediately adjacent to the exact state, as shown in Figure 3.

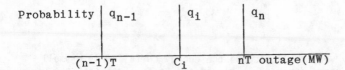

a) Unit representation before rounding

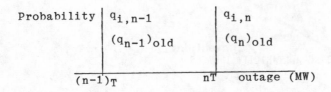

(b) Unit representation after rounding

Figure 3: Unit failure density function.

The sharing procedure is as follows:

$$q_{i,n-1} = \frac{nT - C_i}{T} q_i$$

$$q_{i,n} = \frac{C_i - (n-1)T}{T} q_i$$

After all generating units have been considered and the availability for each rounded state is evaluated, it is possible to delete from the rounded table those states with an availability less than a specified value. Following this point the approximate technique proceeds similarly to the exact technique, discussed in the previous section, to evaluate the COPT.

The use of the rounded table to calculate the risk level introduces certain inaccuracies. The error depends upon the rounding increment chosen and on the shape of the load characteristics and increases with increasing rounding increment (8).

In the two techniques discussed, i.e. the exact method and the rounding method, the number of states in the COPT increases rapidly with increasing number of units. The computation time increases quadratically with increasing number of steps in the rounding technique. It is, therefore, very useful to introduce an approximate technique (9) based on fast Fourier transforms (FFT). The FFT algorithms take advantage of some properties of exponential functions to give fast and precise representation of the random variable in the frequency domain. This particular technique, which has been shown to produce excellent results with a dramatic saving in computation time as shown in Figure 4, is comprehensively discussed in references (8,9,10).

172

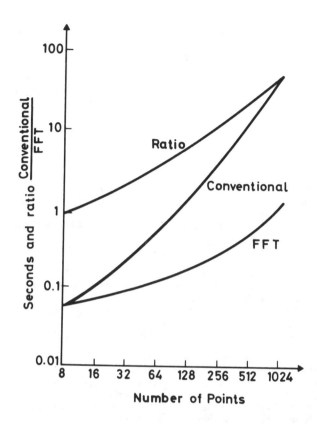

Figure 4: Computational time as a function of number of points.(4)

CONCLUSIONS

The most widely used reliability index in reliability-based system planning is the LOLE, due to its great flexibility and compatibility with alternative expansion plans.

In system planning studies, the decision of how the system should develop is related to reliability, technical and economic reasons. The LOLE index does not answer all the questions and, therefore, other complementary indices such as EENS and EIR have been developed. These provide more information to judge the goodness of a power system and help in making proper decisions. All these techniques have been described in this paper.

In reliability evaluation of large systems, some techniques could become tedious and computationally inefficient. It has been suggested in the paper that another approximate technique based on fast Fourier transforms can be applied to overcome the computational limitations of both the exact and the rounding methods without introducing unacceptable errors in the reliability evaluation process. This becomes increasingly important in the planning and analysis of large modern power systems. This approximate technique is applied in reference 10 and is shown to yield excellent results.

REFERENCES

1. Billinton R, and Allan R N, Reliability Evaluation of Engineering Systems: Concepts and Techniques, pp. 5 - 35, Pitman Advanced Publishing Program, London 1983
2. Billinton R, Power System Reliability Evaluation, pp. 92 - 144, Gordon and Breach, London 1977
3. Billinton R, and Allan R N, Reliability Evaluation of Power Systems, pp 71 - 102, Pitman Advanced Publishing Program, London 1983
4. Sulivan R, Power System Planning, pp. 61 - 94, McGraw Hill, 1977
5. AIEE Working Group Report, 'Application of Probability Methods to Generating Capacity Problems', AIEE Trans. PAS, Vol. 79, pp1165 -1182. pt.III
6. Endreyi J, Reliability Modelling in Electric Power Systems, pp. 114 -143, Wiley International, 1978
7. Takieddine F N, 'Reliability Assessment of Generating Systems', Ph.D. Thesis, University of Manchester, Manchester, 1977
8. Abu-Nasser A, 'Fast Fourier Transform in Generating Capacity Reliability Evaluation', M.Sc. Dissertation, University of Manchester, 1979
9. Allan R N, Abu-Nasser A, 'Discrete Convolution in Power System Reliability, IEEE Transactions on Reliability, Vol. R - 30, No 5, December 1981
10. Shaalan A M, 'Reliability Evaluation in Long-Range Generation Expansion Planning, Ph.D. Thesis, University of Manchester, Manchester, 1984
11. Allan, R and Shaalan, A, 'Cost Benefit and Reliability Assessment of Electrical Generating Systems', Proceedings of National Reliability Conference, Vol 2, 1985.
12. Shaalan, A, 'Reliability Criteria used by Saudi Consolidated Electric Company (SCECOS) in Generating Capacity Planning', Proceedings of the 2nd Saudi Engineers Conference, University of Petroleum and Minerals, Dharan, 16 - 19 November, 1985 (to be published.
13. Ibid, 'Modelling Uncertainty in Long-Range Power System Planning', Simulation and Modelling Conference, Lugano, Switzerland, 24 - 26 June, 1985

12

■

Software Reliability

P. D. T. O'CONNOR
British Aerospace
Hertfordshire, UK

INTRODUCTION

Software is now part of the operating system of a very wide range of products
and systems, and this trend is accelerating with the opportunities presented by
low cost microprocessor devices. In most cases, the fact that computer programs
take over functions previously performed by hardware results in enhanced
reliability, since software does not fail in the way that hardware does.
Performing functions with software leads to less complex and more robust
hardware. Each copy of a computer program is identical to the original, so
failure due to variability cannot occur. Also, software does not degrade,
except in a few special senses, and when it does it is easy to restore it to its
original standard.

Nevertheless, software can fail to perform the function indended, due to
undetected errors in the program. When a software error does exist, it exists
in all copies of the program, and if it is such as to cause failure in certain
circumstances, it will always fail when those circumstances occur. Therefore,
software errors can be extremely serious.

Since most programs consist of very many individual statements and logical
paths, the scope for error is large. A software reliability effort is concerned
with minimizing the existence of errors, by imposing programming disciplines,
checking and testing. The term 'software reliability' is not universally
accepted, since it is argued that reliability implies a probability, whereas a
program either contains one or more errors, in which case the probability of
failure in certain circumstances is unity, or it contains no errors, in which
case the probability of failure is zero. Similarly it is argued that 'rate of
failure' has no meaning in a software sense. A good program will run
indefinitely without failure, and so will all copies of it. A program with an
error will always fail when that part of the program is executed under the error
conditions. However, we will consider the system user's point of view. The
user will observe system failures, and will be equally affected whether they are
caused by hardware failures or by software errors. In some cases it might not
even be possible to distinguish between hardware and software causes.

There are several ways by which hardware and software reliability differ. Some
have already been mentioned. Table 1 lists the differences.

This chapter is extracted from "Practical Reliability Engineering" (2nd edition), Chapter 8, by P. D. T.
O'Connor, with the permission of the publishers, John Wiley & Sons Ltd.

Table 1.   Hardware and software reliability

Hardware	Software
1. Failures can be caused by deficiencies in design, production, use and maintenance.	1. Failures are primarily due to design errors, with production (copying) use and maintenance (excluding corrections) having negligible effect.
2. Failures can be due to wear, or other energy-related phenomena. Sometimes warning is available before failure occurs.	2. There is no wearout phenomenon. Software failures occur without warning.
3. Repairs can be made which might make the equipment more reliable.	3. The only repair is by redesign (reprogramming), which, if it removes the errors and introduces no others, will result in higher reliability.
4. Reliability can depend on burn-in or wearout phenomena, ie. failure rates can be decreasing, constant or increasing with respect to operating time.	4. Reliability is not so dependent. Reliability improvement over time may be affected but this is not an operational time relationship. Rather it is a function of the effort put into detecting and correcting errors.
5. Failure can be related to the passage of operating (or storage) time.	5. Reliability is not time-related in this way. Failures occur when a program step or path which in error is executed.
6. Reliability is related to environmental factors.	6. The external environment does not affect reliability, except insofar as it might affect program inputs.
7. Reliability cannot be predicted in theory from knowledge of design and usage factors.	7. Reliability cannot be predicted from any physical basis, since it entirely depends upon human factors in design. Some a priori approaches have been proposed. (see below)
8. Reliability can sometimes be improved by redundancy.	8. Reliability cannot be improved by redundancy if the parallel program paths are identical, since if one path fails the others will have the same error. It is possible to provide redundancy by having parallel paths, each with different programs written and checked by different teams.
9. Failures can occur to components of a system in a pattern which is to some extent predictable from the stresses on the components and other factors. Reliability critical lists and Pareto analysis of failures are useful techniques.	9. Failures are not usually predictable from analysis of separate statements. Errors are likely to exist randomly throughout the program, and any statements may be in error. Reliability critical lists and Pareto analysis of failures are not appropriate.

SOFTWARE FAILURE MODES

Software errors can arise from the specification, the software system design and from the coding process.

Specification Errors

Typically more than half the errors recorded during software development originate in the specification. Since software is not perceivable in a physical sense, there is little scope for common sense interpretation of ambiguities, inconsistencies or imcomplete statements. Therefore, software specifications must be very carefully developed and reviewed. The software specification must describe fully and accurately the requirements of the program. The program should reflect the requirements exactly. There are no safety margins in software design as in hardware design. For example, if the requirement is to measure 9 ± 0.5 V and to indicate if the voltage is outside these tolerances, the program will do precisely that. If the specification was incorrectly formulated, e.g. if the tolerances were not stated, the out-of-tolerance voltage would be indicated at this point every time the measured voltage varied by a detectable amount from 9 V, whether or not the tolerances were exceeded. Depending on the circumstances this might be an easily detectable error, or it might lead to unnecessary checks and adjustments because the out-of-tolerance indication is believed. This is a relatively simple example. Much more serious errors, such as misunderstanding of the logical requirement of the program, can be written into the specification. This type of error can be much harder to correct, involving considerable reprogramming, and is much more serious in effect.

Software System Design

The software system design follows from the specification. The system design may be a flowchart and would define the program structure, test points, limits, etc. Errors can occur as a result of incorrect interpretation of the specification, or incomplete or incorrect logic.

An important reliability feature of software system design is robustness, the term used to describe the capability of a program to withstand error conditions without serious effect, such as becoming locked in a loop or 'crashing'. The robustness of the program will depend upon the design, since it is at this stage that the paths to be taken by the program under the error conditions are determined.

Software Code Generation

Code generation is a prime source of errors, since a typical program involves a large number of code statements.

Typical errors can be:

1. Typographical errors.
2. Incorrect numerical values, e.g. 0.1 for 0.01.
3. Omissions of symbols, e.g. parentheses.
4. Inclusion of expressions which can become indeterminate, such as division by a value which can become zero.

Structure

Structured programming is an approach that constrains the programmer to using certain clear, well-defined approaches to program design, rather than allowing total freedom to design 'clever' programs which might be complex, difficult to understand or inspect, and prone to error. A major source of error in programs is the use of the GOTO statement for constructs such as loops and branches (decisions). The structured programming approach therefore discourages the use of GOTOs, requiring the use of control structures which have a single entry and a single exit.

Structured programming leads to fewer errors, and to clearer, more easiliy maintained software. On the other hand, structured programs might be less efficient in terms of speed or memory requirements.

Modularity

Modular programmming breaks the program requirement down into separate, smaller program requirements, or modules, each of which can be separately specified, written and tested. The overall problem is thus made easier to understand and this is a very important factor in reducing the scope for error and for easing the task of checking. The separate modules can be written and tested in a shorter time, thus reducing the chances of changes of programmer in mid-stream.

Each module specification must state how the module is to interface with other parts of the program. Thus, all the inputs must be specified. Structured programming might involve more preparatory work in determining the program structure, and in writing module specifications and test requirements. However, like good groundwork in any development programme, this effort is likely to be more than repaid later by reduced overall time spent on program writing and debugging, and it will also result in a program which is easier to understand and to change. The capability of a program to be modified fairly easily can be compared to the maintainability of hardware, and it is often a very important feature. Program changes are necessary when logical corrections have to be made, or when the requirements change, and there are not many software development projects in which these conditions do not arise.

The optimum size of a module depends upon the function of the module and is not solely determined by the number of program elements. The size will usually be determined to some extent by where convenient interfaces can be introduced. As a rule of thumb, modules should not normally exceed 100 separate statements or lines of code in a high level language.

Requirements for Structured and Modular Programming

Major software customers specify the need for programs to be structured and modular, to ensure reliability and maintainability. These disciplined approaches can greatly reduce software development and life cycle costs. In the United Kingdom, MASCOT (modular approach to software construction, operation and test) and, primarily in the United States, ASPE (Ada programming and support environment) are methods imposed to ensure that the great difficulties and costs

generated by the unstructured approaches used in the past are obviated.
Reference 2 in the Bibliography covers structured and modular programming in
more detail.

PROGRAMMING STYLE

Programming Style is an expression used to cover the general approach to program
design and coding. Structured and modular programming are aspects of style.
Other aspects are, for example, the use of REM (remark) statements in the
listing to explain the program, 'defensive' programming in which routines are
included to check for errors, and the use of simple constructs whenever
practicable. Obviously, a disciplined programming style can have a great
influence on software reliability and maintainability, and it is therefore
important that style is covered in software design guides and design reviews,
and in programming training.

FAULT TOLERANCE

Programs can be written so that errors do not cause serious problems or complete
failure of the program. We have mentioned 'robustness' in connection with
program design, and this is an aspect of fault tolerance. A program should be
able to find its way gracefully out of an error condition and indicate the error
source. This can be achieved by programming internal tests, or checks of cycle
time, with a reset and error indication if the set conditions are not met.
Where safety is a factor, it is important that the program sets up safe
conditions when an error occurs. For example, a process controller could be
programmed to set up known safe conditions and indicate a problem, if no output
is generated in two successive program cycle times or if the output value
changes by more than a predetermined amount.

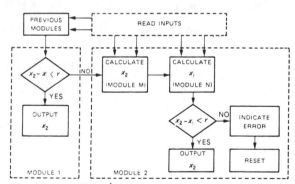

Figure 1 Fault tolerant algorithm

Fault tolerance can also be provided by program redundancy. For high integrity systems separately coded programs can be arranged to run simultaneously on separate but connected controllers, or in a time-sharing mode on one controller. A voting or selection routine can be used to select the output to be used. The effectiveness of this approach is based on the premise that two separately coded programs are very unlikely to contain the same coding errors, but of course this would not provide protection against a specification error. Redundancy can also be provided within a program by arranging that critical outputs are checked by one routine, and if the correct conditions are not present then they are checked by a different routine (Figure 1).

These software techniques can also be used to protect against hardware failures, such as failure of a sensor which provides a program input. For example, failure of a thermostat to switch off a heating supply can be protected against by ensuring that the supply will not remain on for more than a set period, regardless of the thermostat output. This type of facility can be provided much more easily with software than with hardware, at no extra material cost, and therefore the possibility of increasing the reliability and safety of software controlled systems should always be analysed in the specification and design stages.

LANGUAGES

The selection of the computer language to be used can affect the reliability of software. There are two main approaches which can be used:

1. Assembly level programming.
2. High level (or high order) language (HLL or HOL) programming.

Assembly level programs are faster to run and require less memory; therefore, they can be attractive for real-time systems. However, assembly level programming is much more difficult and is much harder to check and to modify. Several types of error which can be made in assembly level programming cannot be made, or are much less likely to be made, in a high level language. Therefore, assembly level programming and the next level down, machine code programming, are not favoured for relatively large programs, though they might be used for modules in order to increase speed and reduce memory requirements.

Assembly and machine code programming are specific to a particular processor, since they are aimed directly at the architecture and operating system.

High level languages, such as BASIC, FORTRAN, Ada, CORAL, etc., are processor-independent, working through a compiler which converts the HLL to that processor's operating system. Therefore, HLLs require more memory (the compiler itself is a large program) and they run more slowly. However, it is much easier to program in HLLs, and the programs are much easier to inspect and correct.

Selection of an HLL also has reliability and maintainability implications, for two main reasons:

1. In some languages, constructs which are frequent sources of errors are more likely to be used. For example, BASIC, while popular and easy to learn, allows GOTO statements to be used for many control structures. GOTO errors are very common. PASCAL and Ada discourage the use of GOTOs, using less error-prone constructs such as IF...THEN...ELSE or DO...WHILE instead.

2. The older HLLs (FORTRAN, BASIC) do not encourage structured programming. PASCAL, Ada and the UK language for defence systems, CORAL 66, strongly encourage structured programming. Ada and CORAL 66 are the standard HLLs in the United States and the United Kingdom for military systems, their use being mandatory in order to ensure system reliability and maintainability.

Since HLLs must work through a compiler, the reliability of the compiler affects system reliability and maintainability. Generally speaking, though, copilers are reliable once fully developed, since they are so universally used. Compilers for new HLLs and/or new procesors sometimes cause problems for the first few years until all errors are corrected.

REAL-TIME SYSTEMS

A real-time system is one in which the software must operate at the speed demanded by the system inputs and outputs. A chess program or a circuit simulation program, for example, will run when executed, and it is not critical exactly how long it takes to complete the run. However, in an operational system such as a process controller or an autopilot, it is essential that the software is ready to accept inputs and completes tasks at the right times. In real-time systems the processor and input and output functions are synchronized by the system clock. The software must be designed so that functions are correctly timed in relation to the system clock pulses.

Timing errors are a common cause of failure, particularly during development, in real-time systems. Timing errors are often difficult to detect, particularly by inspection of code. Timing errors can be caused by hardware faults or by interface problems. However, logic test instruments (logic analysers) can be used to show exactly when and under what conditions system timing errors occur, so that the causes can be pinpointed.

DATA RELIABILITY

Data reliability (or information integrity) is an important aspect of the reliability of software-based systems. When digitally coded data are transmitted, there are two sources of degradation:

1. The data might not be processed in time, so that processing errors are generated. This can arise, for example, if data arrive at a processing point (a 'server', e.g. a microprocesor or a memory address decoder) at a higher rate than the server can process.
2. The data might be corrupted in transmission or in memory by digital bits being lost or inverted, or by spurious bits being added. This can happen if there is noise in the transmission system, e.g. from electromagnetic interference or defects in memory.

System design to eliminate or reduce the incidence of failures due to processing time errors involves the use of queueing theory, applied to the expected rate and pattern of information input, the number and speed of the 'servers', and the queueing disciplines (e.g. first-in first-out (FIFO), last-in first-out (LIFO), etc.). Also, a form of redundancy is used, in which processed data are accepted as being valid only if they are repeated identically at least twice, say, in three cycles. This might involve some reduction in system processing or operating speed.

Data corruption due to transmission or memory defects is checked for and corrected using error detection codes. The simplest and probably best known is the parity bit. An extra bit is added to each data word, so that there will always be an even (or odd) number of ones (even (or odd) parity). If an odd number of ones occurs in a word, the word will be rejected or ignored. More complex error detection codes, which also provide coverage over a larger proportion of possible errors and which also correct errors, are also used. Examples of these are Hamming codes and BCH codes.

PROGRAM CHECKING

Few programs run perfectly the first time they are tested. The scope for error is so large, due to the difficulty that the human mind has in setting up perfectly logical structures, that it is routine for programmers to have to spend a long time debugging a new program until the basic errors are eliminated. Modern high level language compilers contain error detectin, so that many logical, syntactical or other errors are displayed to the programmer, allowing them to be corrected before an attempt is made to load the program. Automatic error correction is also possible in some cases, but this is limited to certain specific types of error. When the program is capable of being run, it is then necessary to confirm that it fulfils all requirements of the specification and that it will run under all anticipated input conditions.

To confirm that the specification is satisfied, the program must be checked against each item of the specification. For example, if a test specification calls for an impedance measurement of 15 1 , only a line-by-check of the program listing is likely to discover an error that calls for a measurement tolerance of +1 , -0 . Program checking can be a tedious process, but it is made much easier if the program is structured into well-specified and understandable modules, so that an independent check can be performed quickly and comprehensively. Like hardware design review procedures, the cost of program checking is usually amply repaid by savings in development time at later stages. The program should be checked in accordance with a prepared plan, which stipulates the tests required to demonstrate specification compliance.

Formal program checking, involving the design team and independent people, is called a structured walkthrough.

FMECAs of Software-based Systems

It is not practicable to perform an FMECA on software, since software 'components' do not fail. The nearest equivalent to an FMECA is a code review, but whenever an error is detected it is corrected so the error source is eliminated. With hardware, however, we cannot eliminate the possibility of, say, transistor failure.

In performing an FMECA of a software-based hardware system it is necessary to consider the failure effects in the context of the operating software, since system behaviour in the event of a hardware failure might be affected by the software. This is particularly the case in systems utilising built-in-test software, or when the software is involved in functions such as switching redundancy, displays, warnings and shut-down.

Testing that a program will operate correctly over the range of input conditions is an essential part of the development process. Software testing must be planned and executed in a disciplined way since, even with the most careful design effort, errors are likely to remain in any reasonably large program, due to the impracticability of finding all errors by checking, as described above. Some features, such as timing, overflow conditions and module interfacing, are not easy to check.

There are limitations to software testing. It is not practicable to test exhaustively a reasonably complex program. The total number of possible paths through a program with n branches and loops is 2^n. It is not normally possible to plan a test strategy which will provide such coverage, and the test time would be exorbitant. Therefore, the tests to be performed must be carefully selected to verify correct operation under the likely range of environments and input conditions, whilst being economical.

The software test process is iterative, whilst code is being produced. It is necessary to test code as soon as it is produced, to ensure that errors can be corrected quickly by the programmer who wrote it. It is also useful to test modules independently, since it is easier to devise effective tests for smaller, well-specified sections of code than for large programs. The detection and correction of errors is also much less expensive early in the development programme. As errors are corrected the software must be re-tested to confirm that the redesign has been effective and has not introduced any other errors. Later, when most or all modules have been written and tested, the complete program must be tested, errors corrected, and retested. Thus design and test proceed in steps, with test results being fed back to the programmers.

It is usual for programmers to test modules or small programs themselves. Given the specification and suitable instructions for conducting and reporting tests, they are usually in the best position to test their own work. Alternatively, programmers might test one another's programs, so that an independent approach is taken. However, testing of larger sections of the program, involving the work of several programmers, must be performed by a separate person or team. This is called integration testing. Integration testing covers module interfaces, and should demonstrate compliance with the system specification.

Formal configuration control should be started when integration testing commences. Formal error reporting (see next section) should also be started at at this stage, if it is not already in operation.

There are two main categories of program testing. Verification is the term used to cover all testing in a development or simulated environment, for example using a host computer. Validation covers testing in the real environment, including running on the target computer, connected to the operational input and output devices. Verification can include module and integration testing. Validation is applicable only to integration testing.

The objectives of software testing are to ensure that the system complies with the requirements and to detect as many errors as is practicable. Therefore the test plan must include:

1. Operation at extreme conditions (timing, input parameter values and rate of change, memory utilisation).
2. Ranges of input sequence.
3. Fault tolerance (error recovery).

Since it is not practicable to test for the complete range of conditions it is important to test for the most critical ones and for combinations of these. Random input conditions, possibly developed from system simulation, should also be used to provide assurance that a wide range of inputs is covered.

ERROR REPORTING

Program

Module

Error conditions:

      Input conditions:

   Description of failure:

Effect importance:

Execution time since last failure: Total run time:

Date: Time: Signed:

Program statement/s involved:

   Line   Statement

Error source:

   Code:
          Design: Specification:

Correction recommended:

   Code:

   Design:

   Specification:

   Date: Signed: Approved:

     Correction made tested: Date: Time: Signed:

     Program master amended: Date: Time: Signed:

Figure 2 Software error reporting form

Reporting of software errors is an important part of the overal program documentation. The person who discovers an error is usually not the programmer or system designer, and therefore, all erors whether discovered during checking, validation or use, need to be written up with full details of program operating conditions at the time. The corrective action report should state the source of the error (specification, design, coding) and describe the changes made. Figure 2 shows an example of a software error reporting form. A software error reporting and corrective action procedure is just as important as a failure reporting system for hardware. The error reports and corrective action details should be retained with the module or program folder as part of the development record.

SOFTWARE RELIABILITY STATISTICS

Efforts to quantify software reliability usually relate to determining (or predicting) the probability of, or quantify of, errors existing in a program. Whilst this is a convenient starting point, there are practical difficulties. The reliability of a program depends not only upon whether or not errors exist but upon the probability that an existing error will affect the output and the nature of the effect. Errors which are very likely to manifest themselves, e.g. those which cause a failure most times the program is run, are likely to be discovered and corrected during the development phase. An error which only causes a failure under very rare or unimportant conditions may not be a reliability problem, but the coding error that caused the total loss of a spacecraft, for example, was a disaster, despite all the previous exhaustive checking and testing.

Error generation, and the discovery and correction of errors, is a function of human capabilities and organization. Therefore, whilst theoretical models based upon program size might be constructed, the derivation of reliability values is likely to be contentious. For example, a well-structured modular program is much easier to check and test, and is less prone to error in the first place, than an unstructured program designed for the same function. A skilled and experienced programming team is less likely to generate errors than one which is less well endowed. A further difficulty in software reliability modelling is the fact that errors can originate in the specification, the design and the coding. With hardware, failure is usually a function of load, strength and time, and whether a weakness is due to the specification, the design or the production process, the physics of failure remain the same. With software, however, specification, design and coding errors are often different in nature, and the probability of their existence depends upon different factors. For example, the number of coding errors might be related to the number of code statements, but the number of specification errors might not have the same relationship. One specification error might lead to a number of separate program errors.

Software that is reliable from the beginning will be cheaper and quicker to develop, so the emphasis must always be to minimize the possibilities of early errors and to eliminate errors before proceeding to the next phase. The essential elements of a software development project to ensure a reliable product are:

1. Specify the requirements completely and in detail.
2. Make sure that all project staff understand the requirements.
3. Check the specification thoroughly. Keep asking 'what if..?
4. Design a structured program and specify each module fully.

5. Check the design and the module specification thoroughly against the system specification.
6. Check written programs for errors, line by line.
7. Plan module and system tests to cover important input combinations, particularly at extreme values.
8. Ensure full recording of all development notes, tests, checks, errors and program changes.

BIBLIOGRAPHY

1. G.J. Myers. Software Reliability: Principles and Practice. Wiley, New York. (1976)
2. M. Shooman, Software Engineering - Design, Reliability, Management. McGraw-Hill, New York. (1983)
3. G.J. Myers.   The Art of Software Testing.   Wiley, New York. (1979)

13
■
Reliability Analysis of Distributed Systems

SALIM HARIRI
Electronic Department
Electrical-Mechanical Engineering
College
Damascus University
Syria

ABSTRACT

Reliability is an increasingly important issue in the design of distributed systems. Conventional reliability measures, such as terminal reliability, are not sufficient to model the reliability of these systems. This paper introduces two new reliability measures, namely, Distributed Program Reliability (DPR) and Distributed System Reliability (DSR). It also presents an efficient approach based on a graph model to evaluate the proposed reliability measures. This method uses breadth-first search techniques to determine all the subgraphs that provide the appropriate accessibility for running a program or a set of programs. These subgraphs are then manipulated by a terminal reliability algorithm to evaluate the reliability.

INTRODUCTION

Distributed systems can be modeled as a collection of different objects that are interconnected via a communication network. In this model, several processing elements cooperate in the execution of a program. For the successful execution of that program, the local host, the processing elements having the required files, and the interconnection links must all be operational. With processing elements and communication links each having a certain reliability, there is a fixed reliability associated with the event in which the program can run successfully.

Several reliability measures have been studied in the context of distributed systems, namely, source-to-multiple-terminal reliability (13), computer network reliability (2) survivability index (11), and multiterminal reliability (7).

The reliability measures mentioned above do not capture the effects of the redundant allocation of a system's resources. However, the two methods described in this paper take redundancy into account. These methods are Distributed Program Reliability (DPR), and Distributed System Reliability (DSR). An elegant approach based on a graph model has been developed to generate all the subgraphs for successful execution of the program(s) under consideration. It avoids applying the path enumeration among pairs of computers, as was done in the multiterminal reliability algorithm (7) and performs a graph traversal to obtain all the required connections represented by trees or forests. These subgraphs are then used to evaluate the reliability.

The organization of this paper is as follows: in the next section, reliability measures for distributed systems are presented. Evaluation algorithms are then introduced to obtain the reliability associated with the proposed measures. An illustrative example is given. Finally, some concluding remarks are provided.

RELIABILITY MEASURES FOR DISTRIBUTED SYSTEMS

Even though many issues, such as distributed algorithms and operating systems, concurrency control, and load balancing have received considerable attention by researchers (3), (4), (5), (14) , there is very little research regarding the reliability of distributed systems. New reliability measures and efficient algorithms for evaluating them are needed to assess the probability that a distributed system with redundant resources meets its specifications.

Consider the simple distributed system shown in Figure 1. Let PAi and FNi denote the set of PEs that can run progrm PRGi and the set of files needed for its execution, respectively (PE denotes a Processing Element or a Computer).

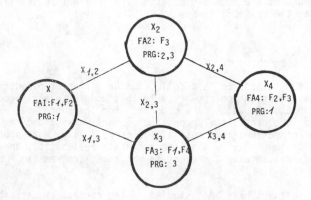

Figure 1.   A distributed system with an allocation of its resources.

Let FAi be the set of files available at node Xi working, and it can access data files F1, F2, F3. In general, the set of nodes and links involved in running a given program and accessing its required files forms a tree. We define a File Spanning Tree (FST) as a tree that connects the root node, where the program runs, to some other nodes such that its vertices hold all the required files for executing that program. A Minimal File Spanning Tree (MFST) is an FST such that

188

there exists no other FST which is a subset of it. For reliability analysis, we are interested in finding all the MFSTs.

For PRG1 to run on either X1 or X4 the MFSTs are:

```
    X1X2X12;        X3X4X34;
X1X2X3X4X13X23    X2X3X4X23X24
```

The probability that a distributed program runs without any failure can be defined as the probability of having at least one MFST operating, i.e.

$$DPR = Pr \left(\bigcup_{j=1}^{nt} MFST_j \right) \qquad\qquad (1)$$

where nt is the number of MFSTs that run a given program.

The DPR measures the reliability of a particular program. For the entire system to be operational, several such programs are required to be operational. We introduce Distributed System Reliability (DSR) as a system level reliability measure. It is defined as the probability of executing m programs successfully, i.e.,

$$DSR = Pr \left(\bigcap_{i=1}^{m} PRGi \right) \qquad\qquad (2)$$

where PRGi denotes the event in which program i is operational.

The subgraph that provides the required accessibility for executing all the programs is called a File Spanning Forest (FSF). The minimum file spanning forest (MFSF) is defined in a way similar to the definition of MFSTs introduced previously.

Equation 2 can be written in terms of the set of all MFSFs as:

$$DSR = Pr \left(\bigcup_{i=1}^{nf} MFSFi \right)$$

where nf is the number of MFSFs that run all programs.

DISTRIBUTED PROGRAM RELIABILITY ALGORITHM

In this section we develop an algorithm to evaluate the reliability of a program based on the graph-theoretic approach. Search techniques can be used to systematically generate all the required trees. The approach of enumerating all the MFSTs is based on traversing the graph in a breadth-first search manner. Once the MFSTs are found, then a terminal reliability algorithm based on path identifiers, such as SYREL1 (8), can be used to determine the DPR. Thus, the algorithm for evaluating the DPR has the following steps:

1. Apply the MFST algorithm to find all the MFSTs for a given program.
2. Apply a terminal reliability algorithm to evaluate the DPR.

MFST Algorithm

In this algorithm the MFSTs are generated in a nondecreasing order of their sizes, where the size is defined as the number of links present in an MFST. First MFSTs of size 0 are enumerated. This occurs when some of the root nodes that run PRGi have all the needed files FNi locally accessible. Next, all the MFSTs of size 1 are determined; these trees have only one edge which connects the root node to some other node, such that the root and the adjacent nodes have all the files of FNi. This procedure is repeated for identifying MFSTs with size 2, and so on up to trees of size n-1, where n is the number of nodes in the system, or no more MFSTs can be obtained.

The procedure to construct the MFSTs consists of checking and expanding steps. In the checking step, trees that have been generated so far will be tested to determine whether or not the vertices of each tree T have all the files FNp needed for executing program PGRp. A tree T is an FST if its vertices have all the required files, i.e.,

$$U \qquad FAi \quad FNp$$
for all Xi   Vt

where Vt denotes the vertices of a tree, say T.

In addition to checking whether or not a tree is an FST, it is also necessary to check if it is an MFST. Once the checking process is completed, a list, say TRY, will have all the trees that are not MFSTs.

The expanding step is necessary to increase the size of each tree in TRY by connecting each vertex of a tree to a new adjacent vertex, hoping that the new vertex will have the needed files. The added adjacent vertex might have all or some of the needed files. The description of the MFST algorithm can be outlined as follows:

Step 1. Initialising. List TRY has the root nodes that run PRGp.
Step 2. Generating all MFSTs.

Repeat

2.1 Checking step. Each tree in TRY is checked to determine whether or not it is an MFST.
2.2 Expanding step. Each tree in TRY is expanded connecting one adjacent edge to its vertices.

Until TRY is empty.

Once the MFSTs have been generated, the next step is to find the probability that at least one of them is up, which means that all the computers and communication links in at least one MFST are operational. Any terminal reliability algorithm based on path enumeration (1), (6), (8) can then be used to determine the distributed program reliability. If the set of MFSTs is considered as a set of paths, the SYREL algorithm, (9) can be used to efficiently evaluate the DPR expression.

EVALUATING DISTRIBUTED SYSTEM RELIABILITY (DSR)

The DSR is defined to provide a global measure and is used to quantify the availability of a certain minimum configuration for the entire system to be considered operational. With our view of a distributed system, this reliability will be the probability that a given set of distributed programs is operational.

Let us assume that a distributed system is defined operational when a set of Np programs are executable in spite of link and node failures. One method of evaluating the DSR is by intersecting the MFSTs associated with each program. The DPR algorithm will be called Np times to obtain the subgraphs that assure the appropriate accessibility for running all the required programs. This approach is simple but computationally expensive, especially when the number of programs involved is large. A detailed description of this method can be found in (12).

A more elegant and efficient approach would be to enumerate directly the subgraphs that will provide the required paths for running all programs. This method is similar to finding the MFSTs, but now it enumerates all Minimal File Spanning Forests (MFSTs). This approach is explained in more detail in (8) and (9).

AN ILLUSTRATIVE EXAMPLE

In this section we illustrate the use of the MFST the algorithm to obtain the DPR for a given program of the distributed system shown in Figure 2. It consists of six processing elements that could run four programs, where each one of them can be executed on two different computers. The allocations of these programs are as follows:

PA1 = (X1, X6);   PA2 = (X3, X4)
PA3 = (X3, X4);   PA4 = (X2, X5)

The set of files needed for executing each program is:

FN1 = (F1,F2,F3);   FN2 = (F2,F4,F6)
FN3 = (F1,F3,F5);   FN4 = (F1,F2,F4,F6)

The files that are accessible at each processing element and the allocation of the programs across the system are shown in Figure 2. In what follows, we briefly evaluate DPR for PRG1. This program can run on either X1 or X6 and its MFSTs shown here are represented only by its edges since they imply the vertices involved in these trees. In the initialization step of the MFST algorithm, the list, say TRY, will have only the root nodes (X1,X6). Then, the checking and expanding steps are repeated until TRY becomes empty. The resulting 10 MFSTs are stored in a list called FOUND and are as follows:

FOUND = (X1,2 X1,3; X1,2 X2,3; X1,3 X2,3; X4,6 X5,6; X4,6
X4,5; X5,6 X4,5; X1,2 X2,4 X4,5; X1,3 X3,5 X4,5 X4,6 X2,4 X2,3;
X5,6 X3,5 X2,3)

Once the MFSTs have been found, the SYREL algorithm (9) can be used to determine the DPR expression. If we assume that all the elements of the system have the same reliability of 0.9, then the reliability of the distributed system for the given set of programs and files equals 0.8419.

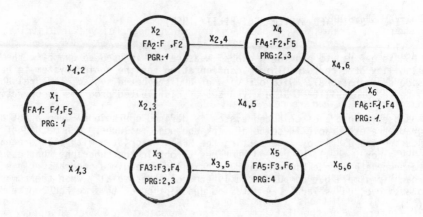

Figure 2: A distributed system with allocation of its four programs and their required data files.

CONCLUSION

In this paper, two new reliability measures, DPR and DSR are introduced to model the reliability of distributed systems. Also, an efficient algorithm for evaluating these measures is developed using a graph-theoretic apprach. These reliability expressions can be used as objective functions that can be used to determine the optimal allocation of a system's resources such that reliability and other performance measures can be optimized.

REFERENCES

1. Abraham J A 'An Improved Algorithm for Network Reliability', IEEE Trans. Reliability, vol. R-28, No 1, pp. 58 -61, April 1979
2. Ball M O, 'Computing Network Reliabiility', Operations Research, vol. 27, No. 4, pp. 823 -838, July - August 1979
3. Berman K P, Joseph, T A, Raeuchle, T, Abbadi, A E, 'Object Management in Distributed Systems', IEEE Trans, Software Engineering, vol. SE-11, No. 6, pp.

502 - 508, June 1985

4.   Chou C K, Abraham J A, 'Load Redistribution under Failure in Distributed Systems', IEEE Trans. Computers, vol. C-32, No. 9, pp. 799 - 808, September 1983

5.   Dion J, 'The Cambridge File Server', ACM Operating System Review, vol. 14, No. 4, pp. 26 - 35, October 1980

6.   Graranov A, Kleinrock L, Gerla M, 'A New Algorithm for Symbolic Reliability Analysis of Computer Communication Networks', Proceedings of the Pacific Telecommunication Conference, January 1980

7.   Graranov A, Gerla M, 'Multiterminal Releability Analysis of Distributed Processing Systems', Proceedings of the 1981 Int, Conf. on Parallel Processing, August 1981

8.   Hariri S, Raghavendra C S, Prasanna Kumar, V K, 'Reliability Measures for Distributed Processing Systems', Proceedings of the Int. Symposium on New Directions in Computing', Trondheim, Norway, August 1985

9.   Hariri S, Raghavendra C S, 'SYREL: A Symbolic Reliability Algorithm based on Path and Cut Set Methods', Proceedings of the IEEE Infocum 86, Fifth Annual Joint Conference of the IEEE Computer and Communication Societies, pp. 203 - 301, Miami, Florida, April 1986

10. Hariri S, Raghavendra C S, Prassanna Kumar V K, 'Reliability Analysis in Distributed Systems', Proceedings of the Sixth Int. Conf. on Distributed Computing Systems, Cambridge, Massachusetts, May 1986

11. Merwin R E, Mirhakak M K, 'Derivation and Use of a Survivability Criterion for DDP Systems', Proceedings of the 1980 National Computer Conference, pp. 139 - 146, May 1980

12. Prasanna Kumar V K, Hariri S, Raghavendra C S, 'Distributed Program Reliability Analysis', IEEE Trans. Software Engineering, vol. SE - 12, No.1, pp. pp.42 - 50, January 1986

13. Satyanarayana A, Hagstrom J N, 'A New Algorithm for Reliability Analysis of Multi-Terminal Networks', IEEE Trans. Reliability vol. R -30, No. 4, pp. 23 -32, October 1981

14. Wittie L, Van Tilborg A M, 'MICROS, A Distributed Operating System for Micronet, A Reconfigurable Computer Network', IEEE Trans, Computers, vol. C -29, No. 12, pp. 1133 -1144, December 1980

14
■
Reliability Testing

R. R. BRITTON
**Plessey Assessment Services Limited
Fareham, Hampshire, UK**

1.   INTRODUCTION

Definitions – types of test.(based on Def.Std. 00–41 (Part 5) Iss.1 Dec.1983,
MIL–STD–785B Sept.1980 and MIL–STD–810D).

Performance – to show that specified performance characteristics are achieved;
no time dependency.

Environmental – to reveal stress dependent weaknesses by subjecting item to
extremes of transport, operational or storage environment. Generally short
duration hence time dependent failures not usually found.

Reliability development – to improve design reliability by sequential
identification, analysis and correction (not just repair) conducted under
'realistic' simulated environments – also referred to as Test, Analyse and Fix
(TAAF) and Reliability Growth (RDGT).

Reliability demonstration (or qualification RQT) – to show that equipment,
representative of approved production standard,achieves specified reliability
requirements with an agreed degree of confidence.

Reliability acceptance (production, PRAT) – to ensure that design reliability
is maintained throughout production by testing samples under highly repeatable
environmental conditions with statistical accept/reject criteria.

Reliability screening – uses functional and environmental stresses (sometimes
unrealistic levels) to induce early life failures due to manufacturing defects
– includes burn-in and environmental stress screening of electronic hardware
(ESSEH). [Reference 1]

Accelerated test – a test designed to shorten test time by increasing the
frequency or duration of environmental stresses that would be expected to
occur during field use.

Aggravated test – a test in which one or more conditions are set at a more
stressful level than the test item will encounter in the field, in order to
reduce test time, reduce sample sizes or assure a margin of safety.

Tailored – the process of choosing or altering test procedures, conditions,
values, tolerances, measures of failure, etc, to simulate or exaggerate the

effects of one or more 'stresses' to which an item will be subjected during its life cycle.

Durability - long term to identify wear-out. It is now generally accepted that environmental and reliability development and qualification tests should be integrated where possible for maximum cost effectiveness.

A properly balanced reliability programme will emphasize growth testing and screening and limit but not eliminate qualification and acceptance tests. See previously quoted documents for further principles of each type of test. These can be applied to all forms of product - commercial, industrial and military.

2.   ENVIRONMENTAL TEST (ET) ENGINEERING

2.1   Scope and Importance

Includes:

a) The study of the individual and combinational environmental stresses imposed upon a product during its life cycle.

b) Derivation of acceptable specifications and practicable test limits.

c) Controlled simulation of these stresses and measurement of the product's response to them - environmental testing.

d) Analysis and understanding of product weaknesses made apparent by such tests.

ET is a means of assessing products for suitability to operate, be stored or transported in a natural man made (induced) environment or combination of both.

Variables are (a) stress level, (b) stress duration. Important stresses include:

vibration	rain
shock	solar radiation
high/low temperature	altitude
humidity	contamination
dust/sand	electromagnetic interference (EMI)

Weaknesses likely to be identified include:

resonances/fatigue	corrosion due to finish/cleaning
loose/weak fixings	leakage of seals
overheating	susceptibility to em. radiation
freezing,seizing,embrittlement	susceptibility to em. conduction

ET is gaining importance because:

a) We are increasingly realising the influences of environment on all living creatures, objects and materials.

b) We now have the means to study these influences and whilst working within tightening cost constraints, our designers are using materials and components closer to their stress limits.

c) As designs become more compact, more automated and sophisticated – the influence of environmental conditions become more significant than for the simple robust objects of the past.

    eg. thermionic valves v semiconductors
          car ignition – magneto v electronics.

d) There is increased transportation for export of goods for installation in other climatic zones than our own.

e) The means are becoming available to simulate the conditions, within very tight control limits under laboratory conditions.

f) Product safety legislation now demands products are tested. It makes economic sense and major procurement agencies demand it.

2.2   Realistic ET Specifications – Test Tailoring

Consider geographic and physical location of item to be tested.
Determine by analysis or measurement the nature and probable incidence of each environmental stress. (Figure 1)

For each relevant stress, consider the complete product life cycle, whether packaged, how mounted, operated, maintained.

Before dealing with specific test methods it is necessary to consider what general types of test can be applied, their objectives and what the criteria are for an acceptable test method.

Simulation

Many environmental tests are simulations of the natural or man made environment. But they are not necessarily faithful reproductions of the actual environment under consideration. Simple reproduction of the natural environment is often ruled out on the basis of test time. Hence many simulation tests are accelerated in some way to compress test time to an acceptable level.

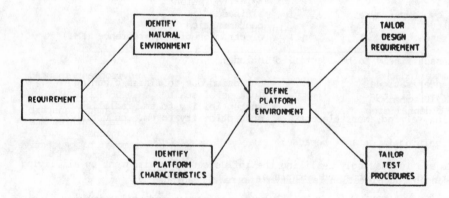

FIGURE 1. Tailoring process

Example 1 – induced environment – vibration and shock

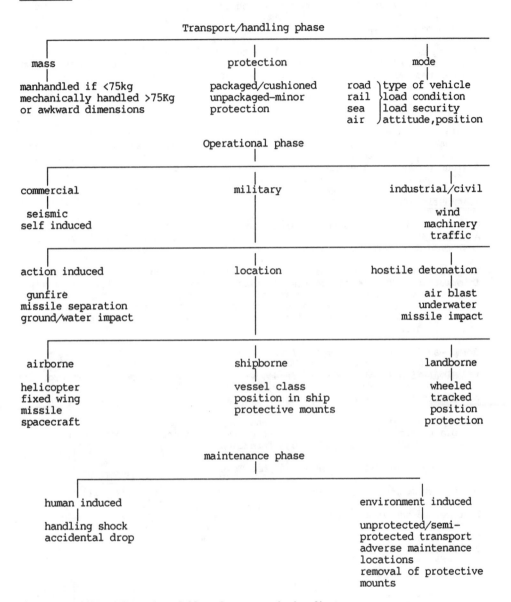

Transport/handling phase

mass	protection	mode
manhandled if <75kg	packaged/cushioned	road ⎫ type of vehicle
mechanically handled >75Kg	unpackaged–minor	rail ⎬ load condition
or awkward dimensions	protection	sea ⎪ load security
		air ⎭ attitude,position

Operational phase

commercial	military	industrial/civil
seismic		wind
self induced		machinery
		traffic

action induced	location	hostile detonation
gunfire		air blast
missile separation		underwater
ground/water impact		missile impact

airborne	shipborne	landborne
helicopter	vessel class	wheeled
fixed wing	position in ship	tracked
missile	protective mounts	position
spacecraft		protection

maintenance phase

human induced	environment induced
handling shock	unprotected/semi-
accidental drop	protected transport
	adverse maintenance
	locations
	removal of protective
	mounts

The means of acceleration fall under two main headings:

a) acceleration by increased stress
b) acceleration of repetitive or cyclic events by increasing either, or both, stress and repetition rate.

197

In the first case the accelerated stress levels are based either on empirical evidence or derived from an understanding of the physical or chemical changes that are being activated. This latter topic is complex and outside the scope of this paper.

Speeding up repetitive events is the alternative method of acceleration. The degree of acceleration attainable will depend upon such factors as duty cycle, thermal mass and time constants, natural resonant frequencies etc.

Arbitrary stress

Several test methods derive from the concept of applying stresses at arbitrary levels to products to provide a 'hurdle' which the design must 'clear'. Although not being correlatable to the real world, such tests, based on sound engineering judgement, provide a consistent standard of performance between various products. The danger of course is that arbitrary stress tests can lead to over design and additional unit cost.

Diagnostic

Both simulation and arbitrary stress tests aim at demonstrating the ability of a product to withstand the environment. They are essentially an 'attribute' test providing a GO/NO GO answer.

Diagnostic tests on the other hand are essentially 'variables' tests providing information of the response or sensitivity of the product to different levels of the environment. Such tests are often carried out at low levels of stress and the results interpreted in terms of the service environment. This extrapolation must be made with care since not all systems will behave in a linear fashion.

Examples of these three types of test are:

1) Simulation,(real time)   – random vibration, shock,soldering drip proof, driving rain
2) Simulation,(accelerated) – Long term damp heat, mould growth, temperature cycling
3) Arbitrary stress   – drop and topple, flammability
4) Diagnostic   – vibration resonance search, oxygen index determination.

2.3   Characteristics of an Acceptable Test

For a test method to be acceptable it must satisfy a number of conditions.

Test method should be:

a) relevant
b) practical
c) repeatable
d) discriminatory.

2.3.1 Relevance

Any test method (even the arbitrary stress type) must relate to the real world in some way. Furthermore the test method must be relevant to the product. It would not be relevant to produce a test method for packaging strength which required attacking the pack with a thermite lance. Nor would it be relevant

to test the resistance to contaminants of IC's by dripping concentrated
sulphuric acid on them. (unless of course part of the production process or
user environment used Conc H$_2$ SO4).

The test of relevance is probably the most crucial and least applied element
of creating a valid test specification.

2.3.2 Practicality

It is important that test methods are practical. If the method is difficult
(or expensive) to set up and operate it should be viewed with suspicion. It
will probably be unrepeatable and unreliable. Practical test methods are
those which do not extend the skills and technology beyond that which is
reasonable today. The use of complicated and esoteric equipment bordering on
the state of the art often produces an impractical test.

2.3.3 Repeatability

Test method specifications usually derive from a National or International
organisation. They are intended for a range of application for a variety of
different products in test laboratories across the world. Confidence in test
laboratories and test methods would be severely eroded if test methods could
not be demonstrated as repeatable from one test to the next or from one
laboratory to another.

Many test methods are not truly repeatable and in some cases there has been
little or no attempt to demonstrate repeatability. This arises for two
reasons.

a)   there are probably more factors determining repeatability than are
     prescribed for checking or calibration eg. turbulent air flow in chambers.

b)   there is no mechanism (in UK) for setting up experiments to evaluate
     repeatability. Test methods are prepared by committees who have no access
     to funds and rely usually on the good offices of one test house for
     evaluation.

2.3.4 Discrimination

The test method must be arranged such that from the outcome of the
test it is easy to discriminate between good and bad. The obvious requirement
is that which is analogous to measurement practice. A 3 digit DVM cannot
discriminate to better than 1% and hence is useless if 0.1% is required. An
effective test method must so stress the product that the effect on a badly
designed product is immediately and effectively apparent.
Badly designed tests often rely too heavily on subjective judgement by the
test engineer as to the outcome or effect of the tests. Such tests are likely
to be poor in discrimination and repeatability.

2.4   Test method format

Bearing in mind the characteristics of an acceptable test method the elements
of the test method specification need considering. These are:

a)   Objective of the test
b)   Test conditions
c)   Stress levels
d)   Procedure

e) Information on criteria of performance
f) Additional information to be given in product specification.

2.4.1 Objective

A clear concise and unambiguous statement is required. There are many examples of poor statements of objective. (eg. BS2011 Test J Mould Growth) even to the extent of (sic).

'The objective of this test is to ascertain whether the product will pass the test'.

2.4.2 Test Conditions

Descriptions of the conditions under which the product will be tested are given, including where required:
- Mounting of the product
- Characteristics of the test equipment
- Control and monitoring methods
- Pre and post test conditioning.

Diagrams and drawings of suitable test equipment and fixtures are often included, together with guidance notes on techniques required for achieving reproducible stress levels.

2.4.3 Stress levels

Many methods provide a general approach to the test supported by preferred severity levels. The objective is to provide a number of different 'standard' stress levels across the total range of the environment. The user is then able to choose a level which most nearly matches his needs. By limiting the number of 'standards' it is easier to make comparisons between the performance of different products.

2.4.4 Procedure

A step by step description of the various phases of the test, including measurements or observations that may be required.

2.4.5 Criteria of performance

To be effective, the test method should clearly indicate the criteria by which the performance of the product during and after the test may be judged.

2.4.6 Additional information

The test method specification being general, does not contain sufficient information to carry out a test. The product specification must therefore provide this data, which may include:

- Pre and post test measurements and performance limits
- Conditioning stress levels
- Operating stress levels
- Duration of test and recovery
- Deviation from procedure, etc.

Test method specifications are available from a number of sources. The most commonly used are:

UK BS2011 (IEC68)   – Basic Environment Testing Procedures
UK BS3G100   – General Requirements for Equipment in Aircraft
UK DEF STAN 07-55   – Environmental Testing of Service Material
US MIL STD 810C   – Environmental Test Methods.

Test methods in these specifications cover a very wide range of environmental
tests. Although each specification may cover similar tests in the other
documents, there are differences in the actual test methods. Appendix 1 gives
the range of tests covered.

2.5   General Points About Testing

Environmental testing often identifies gradual parameter drift rather than
catastrophic failure. If a test takes some time it is often possible that a
set up is broken down or individual instruments replaced with equivalent ones.
It is important that repeatability checks are made or a simple re-calibration
performed. Any observed changes must allow for total measurement uncertainty,
(including measurement resolution).

Measurement uncertainty – reliability growth testing is a stop-start activity.
It is important that measurements of test facilities as well as the test item
are regularly checked against a fixed reference and that test conditions are
highly repeatable eg temperature $\pm 2°C$, relative humidity $\pm 2\%$.

Calibration traceability – the 'pedigree' of each measurement must be known –
particularly if development engineers use one set of equipment and the test
house another.

Remember – measurements can be precise but 'wrong' [Reference 2].

Modern electronic measuring equipment is fast, sensitive and made to measure
and record a wide variety of parameters. Some even have autocalibration
(using a built-in memory), can scan many channels and do scientific
computation but many users do not appreciate measurement fundamentals. Shunt
resistance, stray capacitance, thermal EMF, incorrect VSWR matching, signal
interference and earthing can all corrupt measurements. But, because a
digital number is displayed to several decimal places it is believed.

3.   THE DYNAMIC MECHANICAL ENVIRONMENT – VIBRATION AND SHOCK

3.1   Definitions and Terms – Vibration

Vibration is an oscillating motion about a reference point caused by a
time-varying force which may be periodic eg simple or complex harmonic or
aperiodic, random – see figure 2.

Analysis of complex periodic waves is represented by the Fourier series as the
sum of harmonically related sinusoids.

Note that displacement, velocity and acceleration for simple sine motion can
be determined by successive time differentiation. For complex displacements
differen-tiation of Fourier series significantly changes wave shape of
velocity and acceleration. Once the period is measured and the wave pattern
known, the signal is 'deterministic' ie predictable. Random vibration is
non-deterministic. Standard deviation and spectral density must be determined
by means of probability theory. (Figure 3) [Reference 3]

201

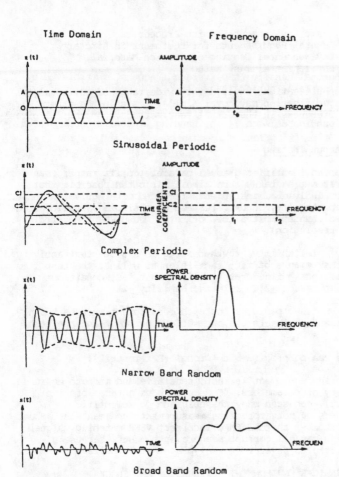

FIGURE 2.   Forms of vibration

The resultant histogram shows relative time the signal exists at different
amplitudes and is a Gaussian distribution. The area under the curve between
two levels is the probability a signal will occur between those levels at any
instant of time. The standard deviation (σ) is a statistical parameter
computed the same as root mean square and can be measured electrically with a
true RMS voltmeter.

In random motion, all frequencies are present all the time. A complex
waveform has a line spectrum ie. all components are distinct frequencies
related to the funda-mental $f_o = 1/T$, where T = period. A random signal has a
continuous spectrum however narrow the portion examined. If a 1Hz bandwidth
is sampled it will still comprise an infinite number of frequencies and
amplitude will vary randomly with time. Spectral density is a measure of
power existing at all frequencies within a band 1Hz wide, usually expressed in
g^2/Hz – see figure 4.

202

Mean square values must be averaged over a series of time samples to obtain a stable value.

Measurement of power spectral density (PSD) versus frequency is achieved by computer calculation using Fast Fourier Transform (FFT) techniques.

Statistical confidence depends upon the number of averages taken and accuracy depends upon 'window' function selected. The Hanning window is usually used for 'stationary' random signals typical of mechanical system vibration (ie. PSD unchanging with time).

A mechanical structure can be considered as an assembly of mass/spring systems with a series of natural frequencies (f_n in Hz) at which modes or standing waves exist. When excited, displacement (d in mm) and acceleration (a in m/sec) are a function of input vector force (F in Newtons), mass (m in kg), stiffness (k in m/kg) and damping ratio (ζ).

Formulae:

$$F = ma \qquad a = \frac{4\pi^2 f^2 d}{1000} \qquad f_n = \frac{1}{2\pi}\sqrt{\frac{k}{m}}$$

$$f_d = f_n\sqrt{(1-\zeta^2)} \qquad \zeta = {}^c/c_c \qquad \text{where } c = \text{actual damping and}$$
$$c_c = \text{critical damping.}$$

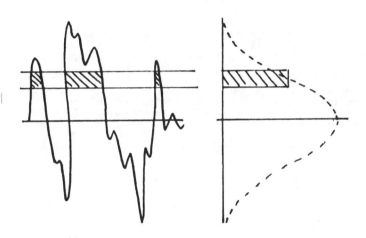

FIGURE 3.   Probability density

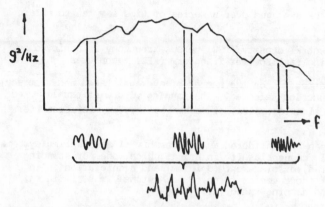

FIGURE 4. Spectral density vs frequency

It is customary to describe the motion in terms of acceleration levels in 'g' units; 1g = 9.81 m/sec/sec.

At resonance, resultant displacement may be many times the input excitation; resultant stresses may exceed yield stress of material or cyclic reversals may cause fatigue.

Typical structures such as electronic equipment, satellites or vehicle bodies are comprised of multiple beams, plates and joints. Resultant motions are complex in amplitude and phase, and may be additive. Response of one item equals input to its connected neighbour.

A structure can vibrate with many degrees of freedom about or through its centre of gravity, often, all at the same time (figure 5).

3.2   Vibration specifications

Sine testing is now generally obsolete except for structural response determination, component testing and rotating machine (eg. helicopter) simulations.

Enveloped wide or narrow band random is now common. Narrow band random or sine superimposed on random is used for tracked vehicle and helicopter simulation. The superimposed sine may be several harmonically related sine waves swept in amplitude and frequency.

Commonly specified (but not necessary actual) levels are shown in figures 6 and 7.

3.3   Test methods and problems

Vibration testing requires:
a) Signal generation – sine or random.
b) Force generator or vibrator with power amplifier
c) Measurement system with signal transducers, amplifiers, filters, servo feedback to generator and output display.
d) Fixture to mount test item to vibration table.
e) Slip table or load bearing platform.

204

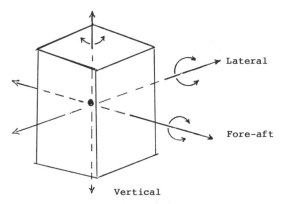

FIGURE 5.   Degress of freedom

Sine generator – basically a low frequency, (3Hz– 10KHz) low distortion (<1%) oscillator with feedback gain control, variable sweep rate (typically 1 octave/minute), high stability (for resonance dwells) and protective overload trips.

Random generators – computer controlled, up to 1600 line resolution (depending on control bandwidth 10Hz – 8KHz). – control error and drive output can be displayed in PSD format with multiple abort or alarm settings. Multichannel monitoring and average/maximum control from multiple inputs is available for complex test items.

Vibrators – basically of 2 types:

a)   electrodynamic – effectively a large loudspeaker system. Amplified drive signal is fed to the armature whilst a high dc current produces magnetic field in the stator coils.

Thermonic valves are still used for the largest amplifiers – poor energy efficiency (<40%) being improved by class D amplifiers.

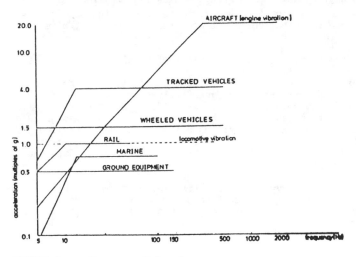

FIGURE 6. -   Degrees of freedom

205

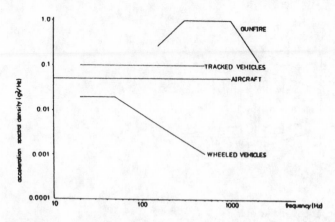

FIGURE 7. Random vibration levels

b) electrohydraulic – oil flow to a ram is modulated through a complex servo
    valve. Not suitable for frequencies above 500Hz due to bulk resonances in
    oil and high phase shift.

    Pneumatic and eccentric mass mechanical systems are sometimes used for low
    cost screening applications.

Measurement system

Accelerometers are basically of 2 types:

a) piezoelectric – figure 8. A seismic mass is pretensioned onto a piezo
    ceramic element; forming a high frequency mass-spring system. When
    acceleration is applied, the oscillatory force produces electric charge
    which is converted by a charge amplifier (typically 50pC/g) and low noise
    cable to a millivolt signal. Shear designs are less prone to sideloads
    (cross-axis), thermal effects and base bending due to mounting/expansion.

b) piezoresistive – semiconductor type strain gauges are mounted on a
    cantilever beam – low resonant frequency but low noise, high sensitivity.
    Some types have integral electronics but most require low drift/offset
    strain gauge amplifiers.

Attachment method of the transducer to the item is critical.

Fixture – the mechanical structure for attaching the test item to the force
input is also critical and must either be dynamically characterised and
representative at all fixing points or be resonance free in the test
bandwidth. This is almost impossible to achieve and most fixture design is a
compromise. Even the torque loading of fixing bolts should be controlled for
repeatability. Performance of the fixture at extremes of test temperature
should be measured. Many misleading vibration (and shock) failures are caused
by incorrect fixturing.

Slip table – used to support heavy items (and fixtures) for vibration in
horizontal planes. Height and alignment of combined centre of gravity are
important design factors as they determine overturning, pitch and yaw motions.

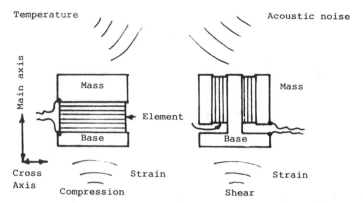

FIGURE 8. Piezoelectric accelerometers

Load bearing platform (or internal load support system) - often required to provide additional support and direction to the armature suspension for heavy items.

Vibration system problems

a) Available thrust - this has to provide required acceleration of the total moving mass - item, fixture, sliptable (and bullnose connector).

b) At resonances the effective moving mass is many times (up to 100) the static mass. Limitation of input force may be necessary or damage to vibrator or item may occur.

c) Test items may be affected by stray magnetic fields. Vibrators should be de-gaussed.

d) Transducers or instrumentation may pick-up unwanted transient noise and cause system to lose control.

e) At low frequencies, (typically <50Hz) and low displacement, due to vibrator travel limitations ($\pm$12mm), accelerometers give low output. Displacement and velocity are difficult to measure as a stationary reference point is required. Many different styles are readily available but care must be taken to select and to apply them correctly. Any transducer should be sensitive to, (and preferably linear with) the particular parameter to be measured and insensitive to all other influences.

f) Measured acceleration will depend upon location of the accelerometer. At any vibration node it will give zero output, so it is important to understand mode shapes. Figure 9 shows first mode shape of a wind turbine blade. The additional mass of the accelerometer may alter the mode properties.

g) Measurement uncertainty for a working vibration system is typically $\pm$7 to 10% and random control is achievable to $\pm$1dB ie. approximately $\pm$ 7% PSD.

h) Availability of sophisticated random test /analysis systems is not matched by the availability of trained dynamics engineers.

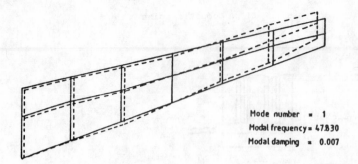

Mode number = 1
Modal frequency = 47.830
Modal damping = 0.007

FIGURE 9. Mode shape of blade

j) The only current way of identifying internal damage to a test item is
either visually or by a change in its functional output. Sensitive
detection of mode shape changes (structural integrity) is being developed
and could of course be applicable following any environmental stress test.

k) Resonance problems can best be overcome by increased damping (cladding,
laminating or friction). Increased frequency by increased stiffness is
related by a square-law. Added mass decreases frequency again.

3.4   Shock Definition and Terms

Shock is a sudden, non periodic disturbance involving large motions in a short
time compared to the natural period of a structure. Few shocks are identical
and shock response spectrum analysis is required to define the amplitude and
frequency content and the equivalent damage potential.

Shock response spectrum (SRS) analysis - if a particular shock waveform
(amplitude v.time) is applied to a structure comprising many individually
tuned, single degree of freedom (ie uniplanar) mass spring systems, (figure
10) each mass will respond in a certain manner. A graph drawn through the
peak positive or negative displacements is the SRS. The response during shock
application is the initial spectrum and that occurring after the input shock
has gone is the residual spectrum.

Classical shock waveforms are used for their known shock spectra,
repeatability and ease of generation.

These are achievable on a computer controlled vibrator or a drop shock machine
using various arresting media or devices. (Figure 11).

Decaying sinusoid pulses are sometimes specified for ship shocks. (Figure 12).

Repetitive shocks are termed bump (unpackaged) or bounce (packaged) and are
produced by closely defined machines.

The simplest form of shock testing is by simple free fall or drop onto a hard
surface. Waveform is unimportant and only drop height and attitude are
specified.   Topple testing comprises releasing an item from its equilibrium
position about a bottom edge and allowing it to fall onto an adjacent face.

208

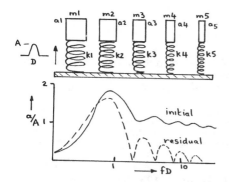

FIGURE 10. Shock response spectrum

3.5   Shock specifications

MIL–STD–810D emphasises that shock tests should attempt to simulate real life measured conditions and recommends the use of shock spectrum synthesis or time history replay. Both require tests to be performed using computer controlled electromagnetic vibrators. Unfortunately, these rarely have sufficient displacement or thrust capability.

3.6   Shock test methods and problems

Low sensitivity, (<1 pico-coulomb/g), high self resonant frequency accelerometers are essential but calibration by direct methods is still a problem.

Monitoring instrumentation must be wide bandwidth (typically 25KHz).

Shock tests are unique events and test equipment and recording system must work first time.

High speed cine' or video is invaluable as actual motion cannot be deciphered from accelerometer responses.

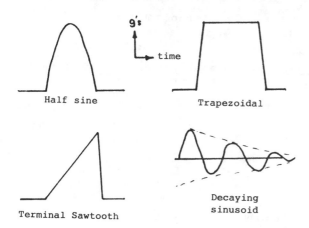

Half sine

Trapezoidal

Terminal Sawtooth

Decaying
sinusoid

FIGURE 11. -   Classical waveforms

4.   THE TEMPERATURE/HUMIDITY ENVIRONMENT

4.1   Definitions and terms

Dry heat and dry cold tests are designed to provide high and low temperature conditions at relative humidity (RH) levels of <20% and with low increase/decrease rates, so that stresses act singly. Typical tests range from –65°C to +125°C.

Relative humidity (%) is the ratio of water vapour mass (g/m^3) present in a given volume of air to that which could be present at saturation conditions at the same temperature – varies with temperature and ambient pressure. Typical test range 20 to 98%.

Diurnal cycling tests are intended to simulate variation and range of temperature conditions due to day and night. Sufficient cycles should occur for large items to acquire equilibrium conditions of temperature and to stabilise their intake of moisture if semi sealed.

Humidity tests may be controlled or specified by dewpoint ie. the temperature at which the vapour pressure of the water vapour in the air is equal to the saturation vapour pressure over water. Generally, RH is measured for environmental testing by means of a psychrometer. The depression of temperature caused by evaporation from the 'wet bulb' of a thermometer relative to an adjacent 'dry bulb' is measured and the equivalent RH is determined from psychometric tables.

Solar radiation comprises the complete ultra–violet, visible and infra–red spectrum of sky radiation scattered by the atmosphere. Measured in watts/metre^2 it produces high surface temperatures and degradation of materials.

4.2   Effects of temperature and humidity

High temperature effects include:

a) primary
   Differential expansion
   Stress relieving
   Softening/melting
   Ageing (of alloys,plastics)
   Setting (of resilient materials)

b) secondary
   High mechanical stress
   Binding of moving parts
   Loss of mechanical strength, distortion
   Loss of adhesion

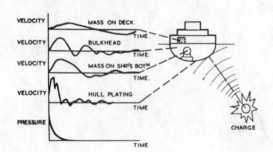

FIGURE 12.   Ship shock – decaying sinusoid

Loss of viscosity
Evaporation/boiling
Increased rate of chemical reaction/
decomposition/diffusion
Change in electrical characteristics

Hardening,embrittlement
Leakage of seals/gaskets
Loss of lubrication
Loss of liquid coolant/
lubricant/etc.
Change (often a deterioration)
in mechanical and chemical
properties and electrical
characteristics

Low temperature effects include:

a) primary

b) secondary

Differential contraction
Hardening of solid materials
Loss of resilience of rubbers, etc.
Freezing of liquids
Contraction/liquifaction of gases
Change in electrical characteristics

High mechanical stress
Embrittlement,cracking, fracture
Penetration of and damage to
seals/gaskets
Seizure through freezing
lubricants
Bursting of items containing
freezing water

Change of temperature effects include:

a) primary

b) secondary

Differential expansion/contraction
Stress relieving
Thermal fatigue

Mechanical stress
Binding of moving parts
Loosening of fixtures
Fatigue failure

High humidity/moisture effects include:

a) primary

b) secondary

Adsorption
Absorption
Erosion (aided by thermal cycling
and mechanical action)
Corrosion (aided by presence of
soluble salts, gases, etc.)
Supports fungal growth

Electrical tracking(short
circuits)
Expansion causing swelling
and changes in physical
properties (eg.breakdown of
organic insulation)
Destruction/disintegration
of material
Chemical/galvanic, stress
corrosion, crevice corrosion,
corrosive fatigue, etc.
Bio-degradation of plastics, etc.

Low humidity effects include:

a) primary

b) secondary

Desiccation
Embrittlement
Static electricity/discharge
(aided by friction of
insulation material)

Shrinkage,changes in physical
properties
Damage/malfunction of electronic
components and circuits)

211

Solar radiation effects include:

a) <u>primary</u>

Infra-red component:high surface
temperatures
Ultra violet component:photo
chemical effects
Visible component:variation
in intensity

b) <u>secondary</u>

Differential expansion, high
temperature deterioration
Deterioration of surfaces/
surface coatings (eg.
embrittlement, crazing, cracking,
discolouration)
Loss of visibility at high and
low light levels

4.3   Test Methods and Practical Precautions

Dependent upon product application, methods may be simple or sophisticated.
Figure 13 shows a simple test on a roof panel exposed sequentially to
simulated solar heating and rainfall. Figure 14 shows a chamber method
necessary for space simulation temperature testing of satellite components.

The method of heat loss from a dissipating test item must be representative in
terms of convection, conduction and radiation. Airflow, wall emissivity and
method of mounting within the test chamber must be considered (particularly
for solar radiation simulation).

Temperature distribution and airflow within a chamber should be calibrated at
several points and a working area defined. Typically, items should not occupy
more than one eighth fraction of chamber volume.

Low temperature test chambers commonly use compressor type refrigeration
systems using R13, R22 refrigerant gases. A cascade system is necessary for
lower than -40°C or to achieve rapid rates of change. Liquid nitrogen (LN_2)
provides a simple solution due to its low equilibrium temperature of -196°C
and cooling capacity of 300 to 400 kJ/kg (dependent on exhaust temperature).
Although expensive to operate, LN_2 is virtually maintenance free and low in
capital cost.

Cold chambers must be sealed or purged to prevent ambient moisture being
sucked in by the reduced vapour pressure and giving ice build up or
condensation 'wetting-out' on the test item. Some tests specifically require
moisture injection, frosting or condensation to simulate aircraft flight
conditions.

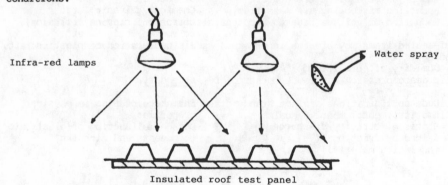

Infra-red lamps

Water spray

Insulated roof test panel

FIGURE 13. Simple solar radiation test

212

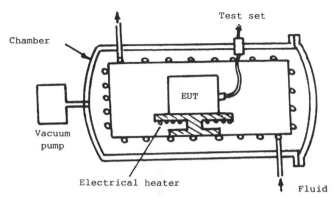

FIGURE 14. Complex solar simulation test

Humidity test chambers require distilled or deionised water to be steam or spray injected and recirculation is not allowed. Conductivity of water <20 micro Siemens and pH range 0.5 to 0.7.

Some chambers work on a bubble-bath principle.

Effects of cyclic humidity are slow so typical test durations are up to 56 and 84 days.

Humidity measurement and control using psychrometers is innaccurate (±5%) as RH approaches saturation because wet-dry bulb differential decreases. Direct RH measurement using electronic (capacitive) sensors is still unproven.

5.   OTHER IMPORTANT CLIMATIC TESTS

5.1   Altitude (low pressure) is obviously important for aircraft items (30,000 metres max.) and many land based applications up to 3,000 metres (eg. Mexico).

Main effects are:

1) Reduction of ambient temperature,
2) Reduction of cooling air mass flow,
3) Reduction of voltage breakdown potentials leading to corona discharge.

Semi-sealed items may breathe in moisture as altitude is reduced eg. aircraft landing.

Rapid de-compression may cause bursting of seals.

Altitude specifications are not standardised and units of measurement still include Torr, mmHg, mbar, Pascals.

5.2   Sand and dust

Effects of sand and dust exposure include:

a) primary

Accumulation on:
- sliding surfaces
- orifices
- points of high static
  potential

Airborne (sand blast) effects

Absorbs moisture etc. to form
'poultice' surfaces (lenses ,dial)

b) secondary

Jamming of threaded devices etc.
Erosion/wear of machine/bearing
surfaces heat exchange fins,etc.
Clogging of air filters,
Ionisation paths formed
Erosion/etching of windows etc)
Severe local corrosion,
electrical breakdown etc.

Specifications such as UK DEF STD 07-55 Test D1 and MIL STD 810D define
grading of particle size, material (eg.silica, red china clay), density
(gm/m^3), velocity (up to 90km/hour) and duration. Correlation of these with
desert conditions is unknown.

Testing the efficiency of engine intake filters for vehicles and helicopters
is critical and requires precise weighing techniques.

MIL-STD-810D shows schematically the elements of a blowing sand/dust facility.
The test item must not occupy more than 50% cross sectional area and
temperature/ humidity of the air must be controlled. MIL 810D now requires
tests at equipment working temperatures. Visual inspection follows for dust
ingress, clogging or sand abrasion. Safety precautions to prevent dust
ingestion are rigorous.

5.3   Driving Rain

Penetration of rain particularly under high wind conditions is surprisingly
common in many exposed equipments. Tests include closely defined spray
patterns at various angles. Opening of the item for visual inspection often
allows accumulated water to ingress further.

5.4   Salt Mist or Fog

Equipment in coastal areas and installed in ships is subject to a humid salt
laden atmosphere which accelerates chemical and galvanic corrosion causing
severe damage. Studies have been undertaken to compare results of standard UK
and US test methods with conditions and effects of 5 months in shore
establishments in Panama and fair correlation achieved. Unfortunately, test
methods have significantly changed and the degree of acceleration or
correlation with real environments is unknown.

Items are mounted in a chamber (20°C) into which an atomized spray at 1% of
chamber volume per hour is injected. The spray mixture (BS2011 Kb) is a
synthetic sea water solution. After 2 hours of spray the items are
stored for 7 days at 35°C, 92% RH. Typically this cycle is repeated four
times.

Electronic equipments and aircraft are a mixture of metallic materials and
form thousands of galvanic 'couples'. Conductive gaskets (EMC) are
particularly prone. Visual detection and interpretaion of corrosion effects
is a major problem.

5.5   Mould or Fungus

This test is applied to materials judged likely to support mould growth.
Considerable health hazard exists to test operators and strict mycological
procedures are essential. Typically, a suspension of nutrient and selected
mould spores is sprayed onto thetest material which is then sorted in a warm
(30°C), humid (>90% RH) environment for up to 84 days.

Mould can cause breakdown of some plastics, adhesives, oils and rubbers. Acid
excretion from moulds causes electrical insulation breakdown and even etching
of glass lenses etc.

5.6   Combined Environments

Natural environments depend upon the geographic location of operation.
Combinations of natural environment obviously occur but are seldom of maximum
magnitude simultaneously eg. maximum temperature with maximum humidity.
Induced environments occur as additional stresses with the natural environment
and are influenced by it.

The total number of combined (dual) environments, of significance, amounts to
more than 80. Combinations which produce an effect greater than the sum of
the effects of each single environment are termed synergistic.

The dominant combinations are mainly those involving:

a) High temperature (softening) or low temperature (embrittlement) and
    mechanical stress such as vibration or shock.
a) Low air pressure and temperature (heating/cooling effects).
c) Corrosive environments with temperature and mechanical stress.

Diurnal variations of the natural environments are also important as also are
sequences of natural or induced environments eg.

a) Rain falling on surfaces immediately after high solar radiation.
b) Cold equipment in unpressurized aircraft bay subjected to saturation on
    landing in high humidity environment.

6.   THE ELECTRICAL AND ELECTROMAGNETIC ENVIRONMENT

6.1   Definitions

Exposure to electrical, electrostatic (ESD) or electromagnetic (EM) stresses
can damage or disturb the functioning of electronic equipments. Tests must
evaluate the degree of protection or immunity to power line transients (spike
and surge), electrostatic discharge and conducted and radiated magnetic or
electromagnetic fields. It is equally important that electronic equipment
does not emit interfering or harmful fields (EMI) either by conduction or
radiation.

Electromagnetic Compatibility (EMC) is the ability of equipments and systems
to co-exist. Generally, tests are applied to prototype or early production
equipment and little is known about degradation of EMC with exposure to other
environments, despite the obvious changes to performance critical materials
caused by mechanical and climatic environments.

215

EM noise can be natural (eg lightning) or man-made. Intentional noise signals (communication) are narrowband to maximise energy in minimum spectrum. Unintentional noise is an unwanted by-product of impulsive currents in rotating machines, switching and digital data systems. Spectrum occupancy is broadband and inversely proportional to pulse width and rise time. Modern digital equipment is particularly susceptible and recent problems include:

a) CB radio affecting petrol dispensers.
b) Auto ignition defects disturbing traffic light sequencers.
c) Tank turret motor controller causing guns to fire.
d) Avionics instruments errors caused by HF transmitters.

The unintentional emission of EM radiation from computers, displays, peripherals and communications equipments can be intercepted for eavesdropping purposes. Potential EMC problems in Computer Integrated Manufacturing (CIM) plants should not be overlooked.

6.2   Test Specifications - divide into two concepts:

a) Industrial - concerned with suppression of emissions that may interfere with TV/radio over limited frequency range 150KHz to 1GHz. The receiver detector is quasi-peak. Output amplitude depends on frequency and duration of interference spikes. Main specifications are CISPR, VDE, UK BS and US FCC and generally require open field site tests.

b) Military - concerned with both emission and susceptibility over wide frequency range 20KHz - 40GHz. True peak detector is used - main specifications are MIL STD 461, 462 and require screened room tests. Hence, specification comparison is impossible unless the nature of interfering signals is known and measurement terminology is totally defined.

Many variables exist so tests must be highly planned and controlled and test sites carefully characterised. Amplitude, frequency, modulation and polarisation of EM signals cannot be applied in all combinations to an equipment which may itself have several different operating modes.

A comprehensive test plan is required based on sound knowledge of the circuit operation within the equipment to be tested. This identifies the relevant subset of the above variables and the test methods (selected from standards). The specified emission or susceptibility limits are broadly determined by the application of the product and its proximity to other electronic equipment.

For military equipment test plans will include:

a) Radiated emission
b) Conducted emission
c) Radiated susceptibility
d) Conducted susceptibility
e) Transient spikes and surges.

Emitted signals must be measured by both narrowband (ie. sharply tunable) and broadband (eg. random noise) methods. Just as in the previous random vibration description, the receiver bandwidth must be stated. Wideband measurements are then normalized to a standard bandwith (typically 1kHz or 1MHz).

216

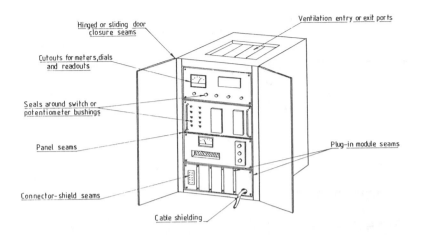

FIGURE 15. Typical shielding problem areas in computer and process control enclosures

6.3 EM Protection

a) <u>Shielding</u> – to contain or exclude EM radiation. Magnetic shielding must be high conductivity or permeability. Electric shielding must be continuous – holes or gaps act as antennae, see figure 15.

b) <u>Filtering</u> – for signal and power cables – effective performance depends on line impedance but is usually only specified at 50Ω.

c) <u>Cables</u> – routing (particularly height above ground plane), twisting, screening – all depend on frequencies present. Installed loop area is critical, see figure 16.

d) <u>Grounding</u> – interfering signals must be returned to source by the lowest impedance path, eg. wide, flat conductors. Earthing arrangements must consider high frequency currents, skin depth, inductance, etc. (c) and (d) will often be changed significantly during production of an item. Techniques apply equally to printed circuit board layouts.

e) <u>Component and circuit selection</u> – use devices and circuit layouts that have a high immunity to EMI. (eg. CMOS digital devices).

6.4 EMC Test Methods/Problems

Because of the many variables listed, EMC testing is highly unrepeatable in both time, equipment sample and between test locations even though signal measurement precision may be high (±2dB discounting the measurement antenna).

Where possible, tests should be performed in a RF quiet enclosure although open field sites are specified for FCC and VDE type testing. Earth plane effectiveness of open field sites may be affected by weather or local buildings can cause reflections. Screened rooms with 100dB of attenuation are used for military tests, fitted with filters for power and signal lines,

217

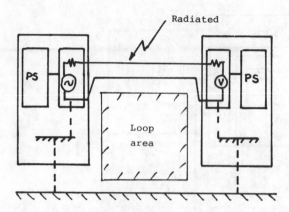

FIGURE 16. Installed loop area of cables

copper bench ground planes, temperature and humidity control and often are anechoically lined to reduce tuned cavity effects.

Figures 17 and 18 show typical screened room complexes designed to segregate the equipment under test (EUT), the test stimulation or ancillary equipment and the EMC measurement or susceptibility generation system. To observe visual signals such as lamps or meters, closed circuit television (CCTV) must be used for susceptibility monitoring as the high RF field strengths are unsafe for humans. Safety interlocks on chamber doors are also essential.

Radiated tests require complex types of antenna, usually large to make them efficient and sited exactly one metre from the equipment under test (EUT). This gives near-field conditions which should not be extrapolated for larger distances.

Conducted tests use clip-on current probes and each conductor in a cable may be separately monitored.Equipment used to stimulate or monitor the EUT must be isolated from it by filtering and shielding. Connecting cable layout must be fixed and repeatable.

Narrowband and broadband tests are performed sequentially for each operating mode of the EUT. Since emissions (or susceptibility) may only occur at a particular point in a program or duty cycle, receivers can only be swept slowly. Testing is time consuming, repetitive and expensive. Computer automated measuring techniques are helpful but cannot better (yet) human recognition of signal anomalies.

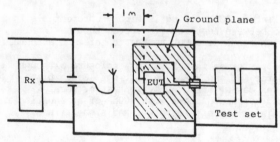

FIGURE 17. Screened room complex – radiated emission

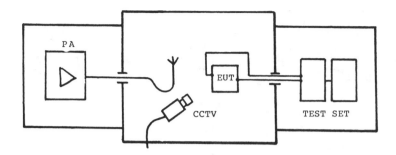

FIGURE 18. Screened room complex – radiated susceptibility

Equipment required for EMC tests:

Emission	Susceptibility
Calibrated receiver(s)	Sweep frequency generator
Spectrum analyser	Power amplifier(s)
Screened room	Screened room
Filtered supplies	Filtered supplies
Antennae (receiving)	Antennae (radiating)
Current probes	Coupling transformers
	Field strength monitor(s)

6.5 Spike and Surge Testing

Completely analogous to mechanical shock, electrical spikes and surges are simulations of transient energy inputs via power or signal lines to an equipment.

There is an infinite variety of real life transients so test transients are closely defined in a wealth of standards and specifications. Again the standards characterise the waveforms not as they occur in nature but rather as they impinge on equipment connected to power line networks. The main standards are IEEE Std. 587–1980 and IEC 664.

Whereas transients on telephone lines would until recently have created an audible nuisance, they can now be disastrous to data. Standard test pulses for signal lines are defined in FCC docket 19528 part 68 and CCITT (Rec.K17).

Impulsive and oscillatory test waveforms are applied either directly or by transformer coupling

7.   COMBINED ENVIRONMENTS FOR RELIABILITY TESTING (CERT)

The primary objective in reliability growth and environmental stress screening (ESS) tests is to promote failures, highlight weaknesses and eliminate the causes. The tests must be carefully defined to achieve this as quickly as possible (expensive to run) without overstress causing unrealistic mechanisms to occur, (often detectable by plotting cumulative failure versus stress level). Combinations of mechanical, thermal cycling and electrical power ON/OFF cycling have proved effective.

219

For reliability growth, conditions and sequences should be substantially based on operational duty cycles but ESS tests should be developed and constantly optimised for maximum effect (ie. stimulation not simulation) - [see Ref. 1].

Mechanical stress (vibration or shock - any axis or all) is best applied at equipment level as it tends to find assembly defects but thermal cycling can be used for components, wiring boards, modules and equipment.

Thermal cycling causes differential expansion and contraction. Obviously, higher rates of change promote larger spatial deviations of temperature and hence mechanical forces at metallic interfaces.

Electrical stress cycling at temperature extremes detects insulation weaknesses, avalanche effects and leakage currents in components but mainly serves to apply additional thermal stresses by local power dissipation.

To detect intermittent or transitory failures, items should be continuously monitored (preferably by data-logger) as the exact combination of causal conditions can often assist analysis.

Manual (or robot) operation of panel controls should also form part of the stress cycle and power supplies should be operated at maximum and minimum specified levels.

7.1 Test Equipment

CERT chambers are expensive. Large heating and cooling plant with rapid air circulation is essential. Note that chamber suppliers quote rates of change of delivered air temperature for a given mass load. The test item temperature will lag. Cycle dwell times should allow for item to stabilise. The effective thermal mass of the test load to be used by the chamber supplier should be defined (eg. surface area).

Vibrators fitted below chambers are thermally isolated and a flexible membrane floor is used to provide vapour sealing. All cable and hand entry ports must be vapour sealed also.

For testing/screening of commercial products, (eg. televisions, micro-computers) thermal and mechanical stress are applied sequentially (no particular order).

Random vibration control can be simplified by allowing lower resolution and spectral/level control accuracy. At the simplest extreme, a controlled impact by a rubber hammer is used.

Typical ESS durations are 12 thermal cycles and 10 minutes random vibration. Large production quantities can be accommodated by a relatively small screening facility.
Growth testing occupies test chambers for long periods and vibrator utilisation is low. Test repeatability is essential so test equipment may be retained for long unused periods whilst failures are analysed and corrected. Additional test time may be achieved by simple module replacement or direct repair of the failed item. This should not delay corrective action but can serve to confirm systematic failures whilst maximising cumulative test time.

8.   RELIABILITY DEMONSTRATION TESTING

The purpose of reliability demonstration testing is to show whether a system (or equipment) has achieved specified (often contractual) reliability level (MTBF).

Samples must be representative of production build and environmental conditions realistic compared with operational use.

MIL STD 781 version B, widely used in the 1970's for avionics, specified unrealistic conditions of rigid thermal cycles, 5°C/minute rate of change, fixed sine vibration at 50 or 25Hz etc. Field reliability levels were significantly lower than demonstrated levels (and predictions). 781C introduced more rigorous environmental conditions of temperature, humidity and vibration based on measured mission profiles and better related to analysis of field failures of avionics equipments due to the following stresses:

Temperature	40%	Sand/dust	8%
Vibration	27%	Salt	4%
Moisture	19%	Altitude	2%
	———	Shock	2%
Sub-total	86%		

For mechanical equipments these ratios will probably vary due to self induced conditions.

781C provides various statistical test plans based on a selected operating characteristic – Figure 19.

In any sampling decision there are risks that the result will not be true. The manufacturer (contractor) requires a test plan to give high probability of acceptance if the measurement is actually higher than the target. The purchaser (consumer) requires a low probability that an actually unacceptable reliability appears acceptable.

Plans are given for producer's risk $\propto$ (range typically 10 to 30%) and consumers risk β (similar range).

Design MTBF Θ_0 should be based on prediction and prior experience. Low limit MTBF Θ_1 is the target to be demonstrated. Design ratio $d = \Theta_0/\Theta_1$.

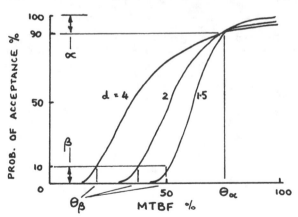

FIGURE 19.   Operating characteristic

221

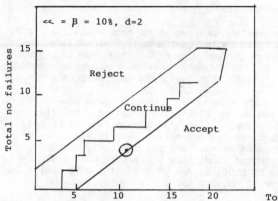

FIGURE 20. Typical PRST plan

The probability ratio sequential test (PRST) plans assume constant failure
rate, therefore samples should be fully screened (burn-in) – but, all
production items must then be given the same burn-in period. Figure 20 shows
failures plotted against cumulative test time (all samples). Accept-reject
decision is made if lines are crossed. Truncation is arbitrariy derived to
limit test time.

A typical test cycle of combined conditions would be as figure 21.

Electrical stress variation typically ± 10%.
Vibration – dependent on application:

a) ground vehicle – swept sine 5–500Hz, ±2g
b) helicopter – swept sine 5–2000Hz
c) jet aircraft – random 20–2000Hz

Thermal – dependent on application but not less than 5°C/min. rate of change.

9.   RELIABILITY GROWTH MONITORING

9.1   Test Philosophy

Reliability prediction is a useful tool for comparison of alternative design
proposals, identifying design areas requiring priority effort or for spares

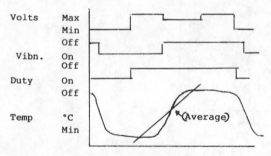

FIGURE 21. Typical test cycle

222

ranging. Input data is dubious and stress models, quality factors etc. unproven.

Demonstration testing and field monitoring generally occurs too late. Burn-in/screening can eliminate or prevent issue of defectively made parts or assembly errors but does not identify intrinsic design problems at an early enough stage.

Therefore it is necessary to employ effective reliability development growth testing using:

a) time oriented tests and realistic environments
b) test analyse and fix (TAAF) approach
c) reliability growth monitoring.

9.2 Growth monitoring

Equipments subjected to a TAAF programme will show decreasing failure rate against accumulated test time well beyond the normal 'burn-in' period - but the actual failures are relatively few. Estimation of the 'current value' MTBF is difficult and subject to large errors.

Several methods have evolved but the Duane method is popular and simple for non statistician engineers/managers to appreciate and communicate to achieve corrective changes.

Duane showed that a log-log plot of cumulative failures rate against accumulated test time produced a declining straight line for systems subject to growth. Assuming exponential distribution, (reasonable for electronic hardware) this can be inverted, to show improvement towards a target by plotting log MTBF θ_c against log time T, see figure 22.

Slope $\propto$ gives a measure of growth rate. Tan $\propto$ = 0.6 is fast growth due to effective TAAF but 0.2 is typical of field failure corrections only.

Formulae: $\log \theta_c = \log \theta_b + \propto (\log T - \log T_0)$
$$\theta_b = \theta_0 \left(\frac{T}{T_0} \right)^{\propto c}$$

At any instant, say after 'n' failures $\theta_c = T/n$ and $\theta_i = \theta_c (1-\propto)$ can be derived.

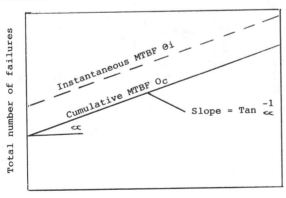

FIGURE 22. Duane plot

The instantaneous MTBF Θ_i is defined as the MTBF that the equipment would exhibit in future if the corrective action programme was stopped at that time and the failure rate remained constant.

The Duane method is applicable to electronic, electro-mechanical, mechanical or hydraulic hardware tests in which a number of samples contribute different amounts to cumulative test time.

Electronic hardware will provide the fastest 'growth' but mechanical or hydraulic equipment will require a test period of two to three times MTBF to exhibit much growth.

Initial results are random and uncertain due to low data amounts and are probably not due to the growth programme effort. Once a recognizable best fit emerges it can be extrapolated and first estimate made of $\propto$.

Management is thus assisted in assessing the likelihood that an equipment will meet its specified target under the test conditions and the expected length of the further testing and corrective action programme can be estimated.

Example

Reliability testing of two identical samples of a communications receiver, using thermal cycling, frosting and bump testing took 16 months. During this time, which included repair and corrective action, 6000 operational hours and approximately 160 defects/incidents were recorded. (This was unusually high but the reason for testing the receiver was that it was unreliable in field use.) Fault classification was:

46% caused equipment to be unable to receive.
33% caused reduced performance.
21% caused no effect on performance.

Further classification showed:

38% – fault not found.
11% – design deficiency.
7.5% – wearout or drift of components.
6% – mechanical damage due to vibration/shock.
5.5% – overheating damage.
2% – due to moisture damage.
5.5% – due to operator/maintainer error.
6% – secondary failures following from other primary events.
13% – due to repetition of non corrected failures.

These proportions did not differ greatly from field experience apart from a higher field incidence of 'fault not found'. (The environmental stress programme rigour probably caused failures to become 'hard' between discrete periods of operational checking).

Several methods of growth measurement were applied as shown in Table 1.

Generally, far fewer failures are available making linear plots and slope estimation less accurate.

9.3   Pareto analysis (significant few, insignificant many)

A Pareto plot of failure data can quickly identify areas requiring priority attention and provide clues to solutions. (see figure 23).

It is common for a majority (>70%) of all failures to be due to only few specific weak areas. If attention is sequentially devoted to these, significant improvements in MTBF will result. Analysis within each category will enable further classification of failure modes and mechanisms and plotting of each failure type with time will show if reliability growth is effected.

10.   CONCLUSIONS

Products, whether for commercial, industrial or mili-tary use must incorporate sufficient protection to with-stand the natural or man made environmental conditions under which they must be transported, stored or operated.

Predictive methods of analysis should be supported by realistic testing.

Table 1. Measurement methods

Method	Θ start hours	Θ_i (hours) (6000)	Θ_i hours (10000)
1. Total failures total time	37	*	*
2. Judgement		54–60	54–60
3. Slope of linear plot		63	>63
4. 2nd order polynomial		192	*
5. Fitted Weibull	36	*	*
6. Duane ($\propto$ = 0.44)	37	66	83
7. Graphical (Crow)		65	81

* = not possible to extrapolate.

Test environments should be logically derived from a rigorous analysis of the product's likelihood of exposure. This will also identify weaknesses in the design specification.

Environmental and reliability development tests should aim to produce as many failures as possible so that weaknesses can be identified and corrected. At realistic test levels this implies:

a) time,
b) sufficient samples,
c) rigorous failure analysis.

Solution of one problem by design modification may cause conflict with another

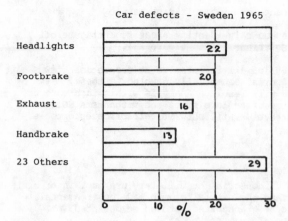

Car defects - Sweden 1965

FIGURE 23 - Pareto plot of car failures

parameter. For example:

a) Sealing for EMC may cause heat dissipation problems.
b) Ruggedising for shock may add weight and reduce resonant frequencies.

The objective of reliability growth analysis techniques is to convince designers, managers and purchasers that corrective action has been effective or more is needed. Simple display methods work best even if not statistically rigorous.

Thermal cycling and random vibration are powerful methods of stress screening, the objective of which is to <u>stimulate</u> failures, not <u>simulate</u> real life.

Formal environmental and reliability testing has mainly evolved from military electronics requirements. The techniques (and benefits) are only now being applied widely to industrial and commercial products.

For ultimate reliability - the KIS principle:
-   keep it simple (minimum complexity, parts count)
-   keep it stable (temperature)
-   keep it still (motion)
-   keep it sealed (ingress)
-   keep it serviced (maintain protection)

What often happens:

-   maximum technical performance and minimum testing cost
-   insufficient thermal and EMC management for real world 'operation'
-   poor dynamic design of equipment practice for mobility requirement, poor packaging etc.
-   weight, materials and cost constraints conflict with access for maintenance
-   maintenance of environmental performance is seldom defined or measured.

226

11. REFERENCES

1. Environmental Stress Screening Guidelines for Assemblies, Institute of Environmental Sciences ESSEH September 1984.

2. Pearce, J.R. — Electronic measuring instruments : results precise but wrong, Electronics and Power Journal IEE pp.155–158, Feb. 1983.

3. Designing electronic equipment for random vibration environments, Proc. Institute of Environmental Sciences, Mar.25–26 1982, Los Angeles, pp.1–6.

APPENDIX 1

BS2011:1977 – Basic environmental testing procedures

Test	Title
Aa	Cold – sudden change of temperature ⎞ non heat
Ab	Cold – gradual change of temperature ⎬ dissipating specimen
Ad	Cold – gradual change of temperature – heat dissipating specimen
Ba	Dry heat – sudden change ⎞ non heat dissipating
Bb	Dry heat – gradual change ⎠ specimen
Bc	Dry heat – sudden change ⎞ heat dissipating
Bd	Dry heat – gradualchange ⎠ specimen
Ca	Damp heat, steady state
Da	Damp heat, accelerated
Db	Damp heat, cyclic
Ea	Shock
Eb	Bump
Ec	Drop and topple
Ed	Free fall
Fc	Vibration (sinusoidal)
Ga	Acceleration, steady state
J	Mould Growth
Ka	Salt mist
Kb	Salt mist
Kc	Sulphur dioxide test for contacts and connections
Kd	Hydrogen sulphide test for contacts and connections
L	(Dust and sand)
M	Low air pressure
Na	Rapid change of temperature, two chamber method
Nb	Change of temperature, one chamber method
Nc	Rapid change of temperature, two water bath method
P	(Flammability)
Qa	Sealing of bushes spindles and gaskets (normal test)
Qb	Sealing of spindles and gaskets (extended test)
Qc	Container sealing, gas leakage (extended test)
Qd	Container sealing, seepage of filling liquid
Qe	Container sealing, penetration of liquid
Qf	Immersion
Qg	Driving rain
Qk	Tracer gas method with mass spectrometer
Ql	Bomb pressure test

```
Sa      Simulated solar radiation at ground level
Ta      Solderability
Tb      Soldering - resistance to soldering heat, Method 1
Va₁     Robustness of terminations - tensile
Va₂     Robustness of terminations - thrust
Vb      Robustness of terminations - bending
Bc      Robustness of terminations - torsion
Vd      Robustness of terminations - torque
A/AD    Composite temperature/humidity cyclic test
Z/AM    Combined cold/low air pressure tests
Z/AMD   Combined sequential cold, low air pressure and damp heat test
Z/BM    Combined dry heat/low air pressure tests
```

BS 3G 100 Part 2 Section 3
General requirements for equipment in aircraft- environmental conditions

Section   Title

3.1   Vibration
3.2   Temperature/pressure
      A. Low temperature, ground survival
      B. Low temperature, ground operation
      C. Low temperature, flight operation
      D. Temperature/humidity
      E. High temperature, ground survival
      F_1. High temperature, ground operation (short term)
      F_2. High temperature, ground operation (long term)
      G. Temperature/pressure low altitude
      H. Temperature/pressure intermediate altitude
      J. Temperature/pressure high altitude
3.3   Mould growth
3.4   Differential pressure
      3.4.3 Excess pressure
      3.4.4 Rapid decompression
      3.4.5 Explosive decompression
3.5   Explosion proof
3.6   Acceleration
3.7   Tropical exposure
3.8   Salt mist
3.9   Ice formation
3.10 Impact icing
3.11 Water proofness
      A. Spray proof
      B. Drip proof
3.12 Fluid contamination
3.13 Resistance to fire
3.14 Acoustic vibration

DEF STAN 07-55 - Environmental testing of service material

Test   Title

A1   Vibration, sinusoidal
A2   Vibration, random
A3   Shock
A4   Drop and topple
A5   Bump
A6   Acceleration (steady state)
A7   Acoustic noise

228

A8	Bounce (wheeled vehicle transportation)
A9	Free fall
A10	Fragility Assessment
A11	Impact (horizontal)
A12	Lifting
A13	Static load (stacking)
A14	Bending
A15	Racking
A16	Test track trial
B1	Constant high temperature, low humidity
B2	High temperature, diurnal cycle, low humidity
B3	High temperature, solar radiation
B4	Constant low temperature
B5	Low temperature, diurnal cycle
B6	High temperature, diurnal cycle, high humidity
B7	Constant high temperature, high humidity
B8	Temperature/humidity, diurnal shelter cycle
B9	Rapid or explosive decompression
B10	Icing/frosting
B11	High temperature/low pressure
B12	Low temperature/low pressure
B13	Low temperature/low pressure/high humidity
B14	Thermal shock
B15	High pressure
B16	High winds
B17	ISAT cycles for explosive items
B18	Temperature/pressure cycles (explosive items)
C1	Mould growth
C2	Corrosion salt
C3	Corrosion acid
C4	Fluid contamination
D1	Dust and sand
D2	Fine mist
D3	Driving rain
D4	Drip proof
D5	Immersion
D6	Sealing (pressure differential)
D7	Snow
E1	Radiation from nuclear explosions
E2	Gamma and X-radiation
E3	Ultra violet, visible light and infra-red radiation

MIL STD 810C – Environmental test methods

Test	Title
500.1	Low pressure (altitude)
501.1	High temperature
502.1	Low temperature
503.1	Temperature shock
504.1	Temperature/altitude
505.1	Solar radiation
506.1	Rain
507.1	Humidity (5 procedures)
508.1	Fungus
509.1	Salt fog
510.1	Dust (fine sand)

MIL STD 810C - Environmental test methods

Test	Title
511.1	Explosive atmosphere
512.1	Leakage (immersion)
513.2	Acceleration
514.2	Vibration (10 procedures)
515.2	Acoustic noise
516.2	Shock (6 procedures)
517.2	Space simulation (7 procedures)
518.1	Temperature/humidity/altitude
519.2	Gunfire vibration, aircraft

15
■
Reliability Prediction

P. D. T. O'CONNOR
British Aerospace
Hertfordshire, UK

FUNDAMENTAL LIMITATIONS OF RELIABILITY PREDICTION

Physical Laws and Models.

In engineering and science we use mathematical models for prediction. For example, electrical power consumption can be predicted using Ohm's Law and the model power = current X potential difference. Likewise, we can predict, with near certainty, future planetary positions using Newton's and Kepler's Laws and our knowledge of present positions, velocities and masses. The laws are valid within the appropriate domain (eg. Ohm's Law does not hold at temperatures near absolute zero or at very high frequencies, and Newton's Laws are not valid at the sub-atomic level). Also, we have no reason to expect that the conditions which ensure the present validity of these laws will change in the future. Therefore we use such laws in physics and engineering with confidence, making appropriate allowance for practical aspects such as measurement errors and other sources of variation.

Whilst most laws in physics can, for practical predictive purposes, be considered as deterministic, the underlying mechanisms of the basic phenomena are often stochastic. For example, the pressure exerted by a gas in an enclosure is a function of the random motions of a very large number of molecules. However, we do not use statistical laws to evaluate pressure, since the enormous number of separate random events and interactions enable us to use the average effect of the molecular kinetic energy, in a simple deterministic and essentially empirical model to predict pressure. Thus Boyle's Law is really a simplification confirmed by observation and by the definitions of the concepts used (pressure, temperature, volume). This is true of other laws, such as Ohm's Law.

Physical situations consisting of a few components have actions and interactions in which the statistical aspects are minimal, for example the laws relating to planetary motion or momentum exchange between snooker balls. Such cases are considered wholly deterministic because of the small numbers involved. However, if we attempt to use these deterministic laws to predict the behaviour of similar but more complex systems, eg. a planetary system consisting of many hundreds of planets or the instantaneous positions of a large number of snooker balls at some time after the first impact, the computational problems become significant and lead to uncertainty of the prediction. Further, minor influences that previously were of no significance, such as the precision of the knowledge of the initial conditions or small external perturbations, become much more significant in relation to the final outcome.

Physical laws are therefore useful predictors of system behaviour either when small numbers of actions and interactions are involved or when very large numbers are involved. In situations where moderately large numbers are involved, the predictive power of science is less certain. This is observable in fields such as aerodynamics and meteorology. Whilst the individual processes are understood and are predictable, the fairly large number of simultaneous interacting processes swamp our capability to compute predicted values using the basic known relationships, even with the use of fast computers.

In practice, the probability that an item or a system will fail, or the rate and pattern of failure, is usually in this area of indeterminacy. Therefore, mathematical models of reliability or failure rate for components, based on factors such as stress, temperature and application, such as have been developed in the past by analysis of failure data and statistical regression analysis, are very unlikely to be good predictors of future behaviour. This uncertainty is further exacerbated by the fact that we often cannot assume that conditions will remain unchanged.

Variable Load and Strength

The probability of failure of an item subjected to a load is a function of the applied load and the resisting strength. Since both load and strength might be variable, the failure probability depends on the way in which the distributed load and strength variations interact over time. In practice, as shown in Chapter 5, predicting the reliability of an item subject to variable loads is highly uncertain, particularly if progressive strength deterioration occurs.

The Human Element

Even if sufficient data were available to overcome the deficiencies of statistical reliability models, an impossible situation in the practical reliability engineering context, the human element can still invalidate any prediction made. For example, a reliability prediction based on failure data during trials might indicate a particular MTBF as likely to be observed in use. However, the prediction cannot effectively allow for the facts that the equipment might be operated and maintained by different people, with varying skills and motivations, or in different environments, or that different quality standards might be applied to future production. Or, for that matter, that reliability achievement might depend upon the equipment being improved in some way, typically by incorporating reliability improvement modifications. Whether or not such improvements are made, and their effectiveness, are under human control, and therefore it is not feasible even to postulate a plausible reliability model.

Failure data, and therefore any reliability model based upon such data, are seriously affected by human intervention and interpretation. For example, in life testing of electronic components, failure might be defined by a parameter being outside a specified value. However, such a parameter excursion will not necessarily be perceived as a failure in an operating system, since the component might be applied in such a way that the parameter excursion does not affect performance. Likewise, perceived failures of systems are often subject to human interpretation, so that what is considered to be a failure by a development test engineer might not be reported as such by an operator, or vice versa.

Summary

Any mathematical model used for reliability predictions is subject to severe credibility limitations, due to:

1. The inappropriateness of mathematical models, as used in physics and engineering, to the domain of reliability.

2. The fact that conditions do not necessarily remain constant over the period of the prediction.

3. Sensitivity to variations of load and strength.

4. Human factors in management, manufacture, application and interpretation.

STANDARD APPROACHES TO RELIABILITY PREDICTION

The most commonly used standard reference for reliability prediction for systems such as military equipment is US Military Standard 756 (Reference 1). It is cited in other standards covering reliability management and methods in defence and non-defence work in the USA and in Europe and elsewhere. MIL-STD-756 covers the following aspects of reliability prediction:

1. The prediction as part of the reliability programme, related to the various phases of the development programme. It requires that an initial reliability prediction is performed as early as possible as part of the feasibility study, to compare options and to validate the design concept. Subsequent predictions are made to update the initial work, using more detailed design information as it becomes available, leading to a final prediction which takes full account of all design and stress information.

2. Procedures for performing the predictions, including modelling methods and data sources. MIL-HDBK-217 is the referenced source for electronic component failure rate data and models (see later).

3. Procedures for documenting and reporting the predictions.

A further useful reference in appropriate circumstances is NASA CR 1129 (Reference 2), which describes the approaches to reliability prediction in the NASA space programmes. Here the stress is laid on reliability prediction as a method for comparing designs, and there is less emphasis on quantitative methods.

A British Standard, BS 5760 (Reference 3), also describes reliability prediction methods and relates them to the reliability programme.

For British defence equipment, UK Defence Standard 00-41 (Reference 4)

includes a section on reliability prediction. Def. Stan 00-41 requires that
MIL-HDBK-217 is used for electronic system reliability prediction. It
includes guidance on interpreting the method for British and European
component standards (eg. BS 9000, CECC) and empirical factors to take account
of the effects of power supply switching and storage, two aspects not covered
by MIL-HDBK-217.

ELECTRONIC SYSTEM RELIABILITY PREDICTION

The most common standard reliability prediction method for electronic systems
is US Military Handbook 217 (Reference 5). A very large database of
electronic equipment and component reliability is used to derive constant
hazard rate models for different component types, based on the Arrhenius
relationship.

The physics of many degradation processes affecting the reliability of
electronic components are such that the process rate can often be described by
the Arrhenius model of chemical reaction rate. As applied to electronic
component reliability, this is:

$$\lambda_b = K \exp \left(\frac{-E}{kT}\right)$$

where λ_b is the process rate (component "base" hazard rate), E is the
activation energy (eV) of the main failure-inducing processes, k is
Boltzmann's Constant (1.38×10^{-23} JK^{-1}, or 8.63×10^{-5} eVK^{-1}), T is the
absolute temperature (K), and K is a constant.

MIL-HDBK-217 in fact quotes "failure rate" models, not "hazard rate" models.
Since most data used for MIL-HDBK-217 are derived from repairable systems, and
since the methods used are mainly applied to such systems, the models really
relate to the failure rate contribution to a system by the component type, ie.
the failure rate of the "socket" into which the component fits. MIL-HDBK-217
assumes constant failure rate distributions for all components, and that the
failure of any component will lead to system failure. It further assumes that
all failures are independent, and that any repair will restore the system to a
"good as new" condition.

These assumptions are rarely valid in practice, as has been shown earlier.
However, MIL-HDBK-217 is still considered to be useful for work in this area
because:

1. A constant failure rate assumption makes system reliability prediction
relatively easy, since an additive (parts count) method can be used. The
implied assumptions of independence of failures and identically distributed
times to failure are considered to be reasonable for most electronic systems,
taking account of the uncertainty of the predicted reliability values.

2. For modelling purposes it is often reasonable to assume that the effects
of maintenance and repair will lead to a tendency to a constant failure rate,
as modules and components are changed. Also, maintenance induces failures
which tend to have a constant rate of occurrence. Therefore there might be an
overall tendency towards a constant failure rate.

3. For logistics planning purposes the constant failure rate model is usually
adequate for prediction and for performance monitoring. MTBF is the
reliability parameter usually used in such cases.

234

4. Alternative models for reliability prediction are themselves subject to such wide margins of error that it is considered that the assumptions behind MIL-HDBK-217 do not make much difference.

The general MIL-HDBK-217 failure rate model is of the form:

$$\lambda_p = \lambda_b \, \pi_Q \, \pi_E \, \pi_A \, ---- \, / \, 10^6 \, h$$

where λ_b is the base failure rate related to temperature (based on the Arrhenius equation), and π_Q, π_E, π_A are factors which take account of part quality level, equipment environment, application stress, etc.

The values of the base failure rates and the various factors in MIL-HDBK-217 are kept up to date by a continuing analysis of failure data on components and systems. However, it must be stressed that failure rate prediction by this method gives only an indication of the performance that might be achieved if there is an adequate reliability and production quality programme.

Limitations of MIL-HDBK-217

The failure rate models in MIL-HDBK-217 are subject to all of the limitations and reservations regarding such models as described earlier in this chapter. However, its use is often a requirement in development contracts, particularly for military equipment. Therefore it is important that users are aware of the general and specific limitations.

The models are based on constant or average stress levels (temperature and electrical stress). They do not take account of variations or of transient overstress, power supply cycling, and other factors that can lead to failure. The models also assume that all devices are prone to failure. This is in practice not a safe assumption, since the only components likely to fail under correct operating conditions are those which are defective in some way, as the great majority of electronic component types have no inherent wearout failure modes. The method therefore takes insufficient account of the effectiveness of modern component quality programmes, and this is a major reason for the often pessimistic system failure rate values it generates.

Experience shows that predictions using MIL-HDBK-217 are typically in the range of 0.2 to 5 times the failure rates achieved by systems, but the errors can be much larger. Therefore the method cannot be considered as a very credible one. In view of the conclusions derived earlier in this chapter this level of uncertainty should not be considered to be in any way surprising.

The failure rate values for microcircuits are particularly pessimistic, mainly due to the improvements in manufacturing quality with the trend towards increasing levels of integration and due to competitive pressures. It seems that the innovations and quality improvements in microelectronics happen too quickly for the authors of reliability prediction methods such as MIL-HDBK-217 to keep pace. This is a good example of the inappropriateness of using a mathematical model to predict the future when the underlying conditions are changing.

Due to the high level of uncertainty associated with each predicted component failure rate, the pattern of failure predicted is very unlikely to match that experienced. Therefore, the predictions cannot be used safely for evaluating component-level criticality or logistics aspects. This is particularly true for microelectronic components, for which the models are very pessimistic, as explained earlier.

A large proportion of the failures of modern electronic systems is caused not by component failures but by other problems such as connectors and connections, miscellaneous quality and design problems, operator or maintainer error, etc. The pattern of such failures obviously varies significantly between types of products and even within a product type. Methods such as MIL-HDBK-217 obviously cannot account for such variation.

Finally, MIL-HDBK-217 is very complex to use, even when computerised. It is at least arguable that the wide margins of error inherent in the method do not justify the use of other than fairly simple formulae and a few key variables. Despite this, MIL-HDBK-217 has become increasingly complex with successive revisions.

Alternative Methods

Other methods have been developed by some organizations for reliability prediction for electronic systems. Examples are those used by British Telecom, Bell Corporation in the USA, and the French telecommunications organization CNET. Some of these methods provide simpler models than MIL-HDBK-217. However, the fact that such alternatives exist, giving very different failure rate values for the same component applications, indicates the uncertainty of the whole process.

The known alternative methods are described in Reference 6.

NON-ELECTRONIC COMPONENTS

Failure rate prediction methods for non-electronic components have not been developed to as complex a position as for electronics. This is mainly due to the much wider range of non-electronic component types and applications. Also, some of the assumptions inherent in the electronic methods, particularly that of constant failure or hazard rate, cannot be justified for most non-electronic components, for which wearout failure modes usually predominate.

Nevertheless, some failure rate data sources have been published, in some cases with factors for environment and application. One of the best known is that produced by the US Department of Defence, Non-electronic Parts Reliability Data (NPRD-2) (Reference 7). However, these methods and data bases are subject to all of the reservations described earlier. Also, there is no generally unit of operating time for many such components as gaskets, springs or optical elements, all listed in such databases. Therefore, such data should be used with at least as much caution as the methods for electronics.

RELIABILITY PREDICTION FOR NON-OPERATING PERIODS

Attempts have been made in the past to provide failure rate prediction methods

236

and data to take account of dormant periods of systems. Several types of system spend much of their lives in non-operating conditions, and the failure mechanisms and rates of the components are obviously different from those in the operating states. For example, a draft version of MIL-HDBK-217 was produced, containing such modifications. However, apart from a rather simpe method given in UK Defence Standard 00-41, no method has gained acceptance, and the latter only within the relevant field. This is to some extent due to the shortage of data on failures due to dormancy; it is not usually easy to tell whether a component failed as a result of doing nothing for a long time or because it was activated. Also, the extrapolation of short-duration inactive accelerated environmental test results over much longer periods is subject to much uncertainty, and most engineering components do not degrade as a result of dormant conditions. Such reliability predictions must be considered therefore to be even more uncertain than those for operating conditions.

RELIABILITY PREDICTIONS BASED ON SYSTEM TEST

Reliability demonstration methods, based on statistical analysis of test results, were described in chapters 3 and 14. These methods assume constant failure rates to justify the statistical basis used. They also make other assumptions which can invalidate the predictions, as described above. The main ones are that future conditions will not change, and that the test environments accurately represent the spectrum of use. Formal reliability demonstration has proved to be as controversial as reliability prediction based on design analysis, and it is a technique which has never found favour in commercial engineering development. It is also now being superseded in military and other system development situations by the more pragmatic approaches to reliability testing described in Chapter 14, in which the objective is to uncover design weaknesses rather than to generate reliability numbers which might be misleading. An example of this more modern approach is described in Chapter 16.

MODERN SYSTEMS

The rapid evolution of microelectronics and software, and new approaches to system design, have further exacerbated the reliability prediction problem. Whilst in the past it was usually reasonable to consider a system failure as having arisen from a component failure, with such modern systems this is not necessarily the case. The increasing complexity and levels of integration of modern systems have created new concepts of unreliability, in addition to the potential for much higher reliability. For example, if failure of a system is defined in terms of a user need (requirement) or a producer's perception of that need (specification), a system can fail in terms of either the requirement or the specification, since the two may not be identical mappings of system operation. In many cases, particularly with microelectronics and software, it is not possible to demonstrate that a system fully complies with its specification. Therefore system failures can be reported which are due not to component failures, but due to specification or design shortcomings. There is no dependable way in which this type of failure can be predicted.

Microelectronics

Even at the component level new problems arise. For example, a microprocessor operates as driven by the system software and by the input data. Therefore, the electrical and thermal stresses encountered at the operating element level are to some extent dependent on the software, which in turn responds to data from the input devices. The same type of microelectronic device in a different application, or being tasked by different software will experience different stress profiles. Electronic reliability prediction methods such as MIL-HDBK-217 make no allowances for such diversity of operation. They view microelectronic devices as simply components, when in fact modern large scale integrated circuit and very large scale integrated circuit (LSI, VLSI) devices are more correctly viewed as sub-systems having orders of complexity many times higher than complete systems being designed only a decade or so ago.

It is not feasible to develop reliability prediction models for separate elements on microelectronic devices, due to the inherent uncertainty as described earlier, and also the extremely high reliability of such elements in practice. In fact, microelectronic devices, and hence their individual elements, must be considered as being inherently reliable when properly manufactured, tested and applied. However, their are some failure modes which can often be considered to be relatively more likely or critical than others, and such relationships can be used in design analyses such as testability analysis or design criticality analysis, in conjunction with the appropriate computer-aided design (CAD) methods. The effects of system software, in terms of how the effects of device failure modes are modified by the software, can also be assessed.

Software

Software is now part of the operating system of a very wide range of products. The trend towards using software for systems control is increasing with the widening application of low-cost microprocessors. In most cases the fact that software takes over functions previously performed by hardware results in considerably enhanced reliability, since, unlike hardware, which can degrade with time, is subject to time/stress relationships, and is subject to variation in production, software is immune to these effects.

A software-driven system can fail when its embedded software contains faults, which are incorrect sections of code due to human inadequacies during specification, design and coding. When such code is executed as a result of certain input data, a processing error results, which, in the absence of fault tolerance, will cause a system failure.

Efforts to quantify software reliability usually relate to predicting or measuring the probability of, or quantity of, faults existing in a program. However, system reliability not only depends on whether or not faults exist in the software, but also on the effects on system operation. Faults which are very likely to manifest themselves, for example those that cause failure most times the program is executed, are likely to be discovered and corrected during the early stages of system development. However, a fault which causes failure only under very rare conditions is more likely to remain undiscovered, but may result in catastrophic failure under certain conditions.

Fault generation during software specification, design and coding, and the discovery and correction of faults during review and test, is a function of human capabilities and organization. Therefore, the derivation of mathematical models of fault count or reliability must be a contentious exercise. For example, a well-structured modular program is easier to check and test, and is thus less prone to be unreliable than is a an unstructured program written to perform the same task. Also, skilled, well-managed staff are more likely to create error-free software. Thus any reliability model must take account not only of system characteristics, but human performance. As explained above, this is not plausible.

Some aspects of software reliability prediction, and a survey of methods that have been proposed, are given in Reference 9.

General

Note that for many modern systems the distinction between hardware and software becomes blurred. For example, the specification, design and test of microelectronic hardware is very similar to the methods used for software, and much of these activities is software-controlled, by CAD and computerised fault emulation and test systems. Therefore reliability prediction for these systems is extremely uncertain, due to the lack of clear boundaries and interfaces, and also due to the extremely high intrinsic reliability.

SYSTEMS RELIABILITY PREDICTION

Methods for evaluating the reliability of complex systems involving redundancy and load sharing was covered in Chapters 4, 10 and 10. It is important, however, that the uncertainty inherent in the reliability values input to the analyses is appreciated and taken into account. Such analyses should include sensitivity studies, to evaluate the effects of large changes in the input assumptions. The size of these variations must be based on experience and on the level of the assessment. For example, the MTBF or availability of a large sub-system can be predicted with less error, assuming that good data are available and it can be assumed that there is unlikely to be significant change, than for a smaller sub-system or a component.

COMPUTERISED RELIABILITY PREDICTION

There is now a wide range of computer programs for reliability prediction, and these should normally be used in preference to manual methods to save time and cost and to enable multiple evaluations to be performed effectively, for example to keep up with design changes or to perform sensitivity analyses.

Several programs are available for performing MIL-HDBK-217 analysis of electronic circuits. These programs range from mainframe systems to low-cost software for personal computers.

Software is also available for reliability modelling work (block diagram analysis, Markov analysis, fault tree analysis).

No prediction of system or component reliability which ignores the effect of production quality assurance can be considered complete or credible. A prediction based solely on design and application considerations, as required by standard methods such as MIL-HDBK-217, ignores the fact that an inherently reliable design can be unreliable because of production quality shortfalls. (MIL-HDBK-217 does not totally ignore quality, but states that it is assumed that appropriate quality standards will be applied). In fact, if the design is inherently reliable and the product is properly used and maintained failures can only be the result of production quality problems. Therefore production quality is often the most important factor in determining product reliability.

The extent to which production quality control approaches perfection, which is the objective of modern approaches to quality control as described in Chapter 17, is a function of human performance and motivation. Therefore it is not deterministically predictable, for the reasons explained earlier. There is no quality problem which cannot in principle be anticipated, circumvented or eliminated. Therefore the effect of quality assurance on reliability depends on the emphasis and management resources applied. In principle, the quality demanded will be the quality achieved, and therefore the reliability prediction must be based on the quality intentions.

Approaches to quality which particularly affect reliability include component selection, vendor control, electronic component screening, test, and control of processes such as assembly and handling. Reliability predictions should therefore take account of how these controls will be applied.

CONCLUSIONS

This chapter has described the fundamental problems involved in attempting to predict the reliability of modern systems. These problems relate to the uncertainty inherent in the physical phenomena associated with failure, and the logical, philosophical and statistical problems associated with attempting to develop credible mathematical models of these processes. The human aspects, in relation to perceptions of failure and human activities that influence reliability, as well as the effects of new technology on the reliability prediction problem have also been described.

Having identified the fundamental limitations of reliability prediction methods and models, we are still left with the problem that reliability prediction is necessary, or at least a useful thing to try to do. Having identified the fundamental limitations of reliability prediction methods and models, we are still left with the problem that reliability prediction is necessary, or at least a good thing to try to do. We also know that it is possible to make reasonably credible reliability predictions under certain circumstances. These are:

1. When there is a known and understood wearout mechanism (physical, chemical, fatigue, etc.), for example the wearout mechanisms of fluourescent tubes or the fatigue life of a gas turbine blade.

2. We believe that the conditions that applied in the past, when the data were collected, will not alter in the future. For example, the fluourescent

tube duty cycles or the gas turbine operation will not be significantly changed, and the future product will be the same as those used in the past.

3. For a new product, the design is similar to those developed, built and used previously, so that our past experience can be applied.

4. The new product does not involve significant technical risks (this follows from 3), or the risks have been identified and brought under control.

4. The system will be manufactured in large quantities, or is complex (ie. it contains a large number of components), or will be used over a long time, or a combination of these conditions exists. Individual prediction errors then tend to cancel out.

6. There is a strong committment to the achievement of the predicted reliability, as an overriding priority.

The above conditions do not all have to apply, but for new designs the last one is essential. Also, it is essential that the predicted reliability is not considered as a ceiling which cannot be exceeded. There is hardly ever a case where it is not practicable and cost-effective to achieve higher levels of quality and reliability by improving designs and production quality standards.

Following these principles, we can make credible reliability predictions for a new TV receiver or automobile engine. No great changes from past practice or utilisation will be involved, technical risks are low, the products are quite complex and will be built in large quantities, and in present markets they must compete with established, reliable products, so the motivation will exist. However, for a new, high technology defence system many of these conditions might not apply, though nowadays the commitment to reliability is generally stronger than in the past.

Note that the reliability predictions for the TV set or the automobile engine could be made without recourse to mathematical models or component reliability data bases. They could be based on knowledge of past performance at the system level and on management targets and priorities. In fact this is usually a more logical and realistic approach.

A reliability prediction does not ensure that the reliability values will be achieved. It is not a demonstration in the way that a mass or power consumption prediction, being based on physical laws, would be. Rather, it should be used as one input to setting the reliability objective, which in turn is likely to be attained only if the right committment exists. This was well illustrated by the notable reliability achievements of the US space programme and the Japanese consumer products industry, neither of whom placed any credence in the standard reliability prediction approaches.

Reliability prediction for new, high technology products must, therefore, be based upon identification of objectives and assessment of risks, in that order. This must be an iterative procedure, since objectives and risks must be balanced. It must also be fully integrated with other design assessments, such as cost, produceability, etc. The reliability engineer plays an important part in this process, since he must assess whether objectives are realistic in relation to the risks. This assessment can be aided by the educated use of appropriate models and data, which help to quantify the risks. For example, it would not be prudent to predict a reliability value greatly in excess of previous experience in the absence of a very strong commitment to achieving it, and the data can help to quantify this. Once the risks have been assessed and the objective quantified, development must be continuously monitored in relation to the reduction of risks through analysis and tests. This is necessary to provide assurance that the objective will be met, if need be by additional management action such as provision of extra resources to solve particular problems. It might even be necessary to change the objective, and intelligent use of data and methods can help the reliability engineer to predict future performance. Intelligent use implies that he understands the actual problems involved in the development programme, knows the action that will be taken and its likely effectiveness, and understands the limitations and uncertainty of the prediction methods used.

Finally, the reliability prediction must be based on the degree of committment towards achieving the reliability objective. This is particularly important in relation to the production phase, when manufacturing quality control becomes the major determinant of in-use reliability, assuming that by this stage the design is correct. (Reference 10.)

REFERENCES

1. US Military Standard 756, Reliability Prediction.

2. NASA CR-1129, Reliability Prediction.

3. British Standard 5760, Reliability of Systems, Equipments and Components.

4. UK Defence Standard 00-41, Practices and Procedures for Reliability and Maintainability.

5. US Military Handbook 217, Reliability Prediction for Electronic Systems.

6. O'Connor, P.D.T. and Harris, L.N., Reliability Prediction, a State of the Art Report, Institution of Electrical Engineers, London, 1986.

7. Non-electronic Parts Reliability Data (NPRD-2), US Department of Defence Rome Air Development Center.

9. Dale, C.J. and Harris, L.N., Approaches to Software Reliability Prediction, Proc. Annual Reliability and Maintainability Symposium, IEEE, 1982.

10. O'Connor, P.D.T., Reliability, Measurement or Management?, Proc. UK National Reliability Conference, Institute of Quality Assurance, 1985.

16
■
Weapon System Reliability Program: A Case History

P. D. T. O'CONNOR
British Aerospace
Hertfordshire, UK

INTRODUCTION

This chapter describes the reliability programme applied to the evolutionary development of an inservice weapon system. It concentrates on the reliability test aspects.

EVOLUTIONARY DEVELOPMENT

In commom with many systems developed more than 15 years ago, the system did not have the benefit of a formal reliability programme. In order to achieve improved reliability in service, many modifications were introduced to in-use equipment. These have generated steady reliability growth.

The evolutionary development activity involved further reliability and performance improvements, including:

1. Hard-wired digital electronics replaced by distributed microprocessors on MIL-STD-1553 data bus.
2. Full built in test equipment (BITE) for system confidence tests and fault diagnosis.
3. Improved performance.

RELIABILITY PROGRAMME

A formal reliability programme was applied for the first time to the system during the recent development programme, related to all new and modified designs and to the overall system. The programme was based on the requirements of UK DEF STANS 00-40/41. The programme comprised the following activities:

1. Reliability prediction.
2. Failure mode effects and criticality analysis.
3. Reliability growth testing.
4. Reliability programme management.

These activities are discussed in detail below.

This chapter is derived from a paper published in the proceedings of the 1985 Annual Conference of the US Institute of Environmental Sciences.

RELIABILITY PREDICTION

Reliability predictions were performed on all new and modified designs. MIL-HDBK-217D was used for all electronic circuit stress analysis reliability prediction, as this method is called up in UK DEF-STAN 00-41. The method provided a convenient independent electrical and thermal stress analysis, as well as an opportunity for design rule checking on aspects such as derating and protection. For non-electronic parts a combination of data obtained from previous customer use of the system, tailored by the expected effectiveness of design changes, and other data sources such as RADC NPRD-2 was used. The objective of the reliability prediction work was to highlight potential critical areas and to recommend design changes, rather than to generate numbers. Therefore, the engineers involved in this work, as with all other aspects of the programme, were thoroughly familiar with the design and operation of the system.

The PREDICTOR reliability program was used to automate all the MIL-HDBK-217D work, allowing the engineers to concentrate on engineering aspects and to keep up with design changes and to evaluate the effects of changes in other inputs such as thermal data and parts specification changes.

Figure 1 shows the results of the reliability prediction of one of the redesigned sub-systems.

FAILURE MODE, EFFECTS AND CRITICALITY ANALYSIS (FMECA)

Detailed FMECAs were performed on new and changed designs. The PREDICTOR FMECA program was used for this work, enabling careful yet econmical analyses to be carried out iteratively as the design evolved. The FMECAs were directed at operational effects of failure, as well as BITE effectiveness and safety aspects. FMECAs were performed at system and sub-system level, and the analysis extended downwards to part level for new designs. The FMECAs proved to be extremely valuable in generating design changes sufficiently early to enable change costs to be kept low. They were also very useful for BITE evaluation and for aiding the preparation of test and diagnostic routines. The combined use of experienced maintenance engineers and CAD R software enabled us to produce timely, effective analyses at low cost.

Figure 2 and Table 1 show the results of the FMECA on one of the sub-systems.

RELIABILITY GROWTH TESTING

All new and redesigned sub-systems were subjected to a reliability growth test (RGT) programme. The objectives of the RGT were to highlight reliability problems which might have been overlooked in the design analyses, and to validate the ability of the system to meet its reliability requirements in service. The test philosophy was essentially a test, analyse and fix (TAAF) approach in accordance with the principles of MIL-STD-1635. Numerical analysis was a secondary objective, and the tests were specifically planned not to be reliability demonstration test of the MIL-STD-781 type. Table 2 shows the pattern of failures generated by some of the sub-system RGTs. Figures 3 and 4 show the pattern of failures on one of the sub-systems, as generated and the expected failure distribution after design corrective action on the failure modes identified by the RGT. Figure 5 shows the MTBF growth during the RGT on this unit, as measured and as adjusted to take account of the effectiveness of corrective action.

In addition to generating information on which to base design corrective action, the RGT was used to validate the design thermal analysis.

The RGTs were carried out in environmental test chambers, programmed to simulate the worst case thermal and operational cycling. For some sub-systems the RGT included extensive cross country transport, to provide a thoroughly realistic and cost effective test profile. The RGTs included user confidence tests using the system BITE wherever practicable.

The RGT was automated to a large extent by using a Hewlett-Packard HP9816 computer to control the units under test, in conjunction with a LORAL SBA100A data bus analyser. The DBA enabled the rest of the system, including the system main computer, to be simulated, using system operational software. The data bus traffic was continuously monitored and stored on a Winchester disc high-speed memory. In the event of failure of the equipment under test, the system ensured that the data bus traffic for the period immediately preceding the failure was captured and the test stopped. This facility was of great value in failure diagnosis.

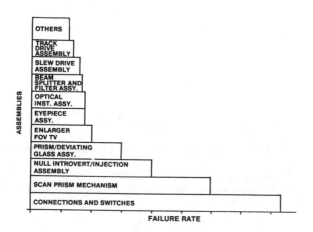

Figure 1.   Subsystem Reliability Prediction

The HP9816 system was also used for RGT failure data collection and analysis, including the preparation of daily logs, individual failure report preparation on the screen, and report preparation including graphics presentations. By using the computer to assist in both test operation and mangement the whole test programme was greatly facilitated.

CONCLUSIONS

The main conclusions drawn from the reliability programme were:

1. It is essential that reliability engineers are familiar with the design and

Description	Failure Mode	Mission Effect	Percent Contrib.	Method of Detection
Plugs and Sockets	Open Circuit	No data/Power transmission	8.5	Not unambiguously detected
Scan Mechanism	Elevation tacho failure	No data transmission	6.7	Bite Test 17
Prism/deviating glass assy.	Misaligned optics	Optical paths not parallel at TV Narrow	5.3	Bite Collimation Test 7
Optical Inst. Assy.	Misaligned optics	Optical paths not parallel at TV Narrow	4.2	Bite Collimation Test 7
Plugs and Sockets	Short circuit	No data/power trans-mission- consequential damage	4.2	Bite Not unambiguously detected
Plugs and Sockets	Intermittent continuity	Intermittent data /power transmission	4.2	Bite Not unambiguously detected
Plugs and Sockets	Physical damage	No data/power transmission	4.2	Obvious interface damage on LRU removal
Scan Mechanism	Misaligned optics	Optical paths not parallel	4.2	Bite Collimation Test 7
Null Introvert/ Injection Assy.	Mechanical misalignment	Optical failure	4.0	Not detected

Table 1.   Sub system FMECA

1.

Total failures = 27
9 Random
7 Manufacture/Procurement
8 Design
2 Testgear
1 Accidental Damage

2.

Total failures = 13
1 Random
7 Manufacture/Procurement
5 Design

3.

Total failures = 60
No Random
48 Manufacture/Procurement
12 Design

Table 2.   RGT Results on 3 Subsystems

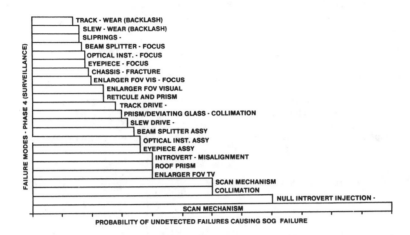

Figure 2. Subsystem FMECA: BITE analysis

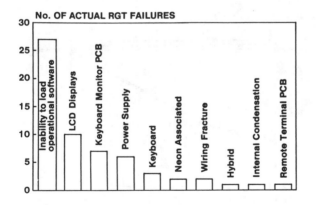

Figure 3. Subsystem RGT Results

with the way the system is operated and maintained. Reliability engineering, to be effective, must not be allowed to be an 'off line' activity.

2. It is important that reliability prediction and FMECA be realistic engineering analyses, performed independently of design but in full partnership. Realistic design analysis, which generates useful outputs for designers, ensures that the reliability engineers are highly motivated and are respected by their colleagues in design.

3. Top mangement motivation and direction is essential. Both customer (UK Ministry of Defence) and Company management were enthusiastic in initiating and supporting the reliability programme.

4. Reliability growth testing is a much better and more cost effective way of improving product reliability than is reliability demonstration testing. The test objective becomes one of reliability improvement, so equipment failures are considered to be test successes, pointing the way to product improvement. Motivation and customer-supplier relationship is greatly improved, since there is no attempt to hide failures or to argue over failure classification.

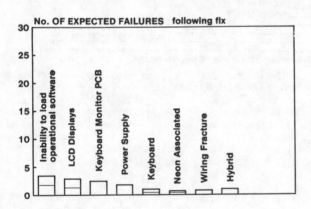

Figure 4. Subsystem RGT: Post-fix Expectation

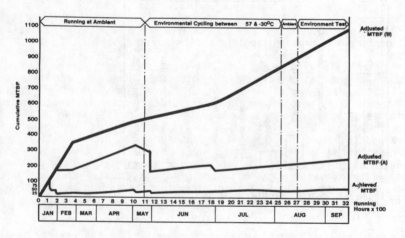

Figure 5. Subsystem RGT: MTBF Growth

248

17

Production Quality Control

DAVID HUTCHINS
David Hutchins Associates
Berkshire, UK

PRODUCTION QUALITY CONTROL

INTRODUCTION

First of all, to clarify points made later, it is necessary to define the terms used in the title of this paper.

The term "Production" is usually understood to infer "engineering production". However, in the context of this paper, the word is intended to infer anything which is either "made", "altered", or "added to" as part of a regular business activity. It includes clerical operations, data processing, in addition to the entire spectrum of product manufacture including food, pharmaceutical, electronics, engineering, printing, chemical, etc.

The term "Quality", when used in an industrial context is usually given to infer quality in terms of defects, errors or faults. In this paper, a broader definition is intended. It includes anything which can be regarded as quality related in any situation, ie. "quality of organisation", "quality of training", "quality of service", "quality of housekeeping", etc, in fact anything which could be used either tangeably or subjectively as a means of judging the attitudes, performance, or conditions in the given environment.

The word "Control" is also subject to different interpretations. "Control" implies that we are satisfied with the Standards achieved, and therefore refers to the means by which we intend to maintain that level of quality e.g. "to provide early warning and the means of corrective action to prevent adverse changes". In other words as a means to "control" variability. Most of the teachings on the topic imply an assumption that variability is inevitable, and therefore the purpose of controls is to prevent matters from getting worse. This is an extremely dangerous fallacy. It is dangerous because not everyone sees it this way. The Japanese for example vigorously challenge the inevitability of variability. "Control" to the Japanese does not infer maintenance of the status quo. Control techniques are used not simply as a means of holding current standards but as tools for continuous project by project improvements.

Anyone who uses quality control just as a means of holding quality levels runs the risk of being left a long way behind by those who adopt the latter approach. In this paper, the word "control" should be taken to be synonymous with "improvement" and these terms will be used interchangeably in this paper.

249

'Quality Control' therefore should be seen as part of the means by which we attempt to harness all the resources available to an organisation to become or continue to be, the best in its business, whatever that may be.

STATE OF THE ART - PAST, PRESENT, FUTURE?

No aspect of Management Science has evolved more extensively than the Science of quality during the past 8 decades, and this evolutionary change is currently taking place at an increasing pace. Perceptions which were regarded as absolute only a few years ago, are currently being totally reappraised. Prior to the industrial revolution, all products were made by people referred to as craftsmen, and a craftsman by definition is responsible for the quality of his or her work. It is of course well known that the craftsmanship system was largely maintained in Europe through the first 100 years of industrialisation by the establishment of the Craft Guilds, Indentured apprenticeships and so on. However, the system began its decline at the turn of the century, as a result of circumstances which prevailed in the United States.

There, the growth of industry outstripped the supply of skilled craftsmen. Fredrick Taylor, a Mechanical Engineer, and other contemporaries suggested that planning and control activities should be separated from the performance of the tasks. Management should "manage" and people "do". To achieve this, tasks were deskilled by breaking them down to their smallest possible elements, and all problem solving elevated to Management and Management Specialists.

This lead to the need for the inspection of work to be carried out by others than those performing the tasks. Inspection as an "on line" management initiated activity began in the first two decades of this century. Soon afterwards, Walter Shewhart evolved the means by which statistical techniques could be used to provide early warning of adverse trends for control purposes.

In the immediate post war era, Statistical Quality Control as it was known was seen as the complete answer to quality related problems. At the same time, large procurement bodies, in particular those in the military field and in automobiles began to insist that their suppliers should operate inspection activities on a systematic and organised basis.

In the 1970's this requirement was superseded by the demand that suppliers should operate quality systems throughout the entire operation.

While this approach has been partially successful, it has not been without its problems. For example, one of the requirements of the procurement bodies, has been the insistence that their suppliers operate the process right down through the entire supply cascade. This has resulted in some companies, which are unfortunate enough to sell their products to a large number of customers, having to face invasions of assessment teams. One British Company claimed that over 80 assessment teams had visited them in a six month period!

Another problem, in the view of the author is the impression that the advantages offered by this approach have been grossly oversold. For example, in the UK, a British Standard for Quality Systems has been produced.(B.S.5750) One Company, given unqualified approval to this standard, admits that over 40% of its product is consistently rejected by the customer and over 80% reworked internally before it ever leaves the factory.

The reason for the failure which is now being recognised is the simple fact that whilst quality assurance and quality control are important, quality systems by themselves do not motivate people. If people do not care, if they just do their jobs according to instructions, then no amount of quality control will give that company the sparkle that separates them from their competitors. Quality control does not hold the pencil of the designer or turn the handles on the machines.

Equally, many of the problems which are endemic in most organisations exist not through incompetence but through the form of organisations we have created based upon the Taylor System. Departmental goals are put before company goals. This frequently gives rise to situations where solutions to problems are found in one department which only makes things worse elsewhere. This problem can be resolved through teamwork. Interdepartmental teams of Managers and Specialists trained in problem solving skills can help to convert an atmosphere of blame into one of collaboration. Or, as Professor Ishikawa, the Japanese quality expert, puts it, "to develop the horizontal fibres of organisation". In the West, we have tended only to develop the vertical fibres. This realisation leads to the current state of awareness.

Analysis of the phenomenal recent success of the Japanese has shown that the goals of quality can only be achieved through a co-ordinated company wide approach. If this concept is universally applicable, and recent experiences indicate that it is, then clearly, we cannot view individual elements such as "production quality control" in isolation from the other elements.

During the current decade, there is widespread realisation amongst the world's leading companies that the application of Quality Science must be rethought. The quality systems have their place, but can only work effectively when proper attention is also given to:

Process Control/Improvement,
Management Style & Organisation,
People involvement,

This paper looks at each of these.

VARIATION CONTROL METHODS

Relationship between Design & Production

In the 1960's variation was regarded by production departments as a problem for production specialists to handle. In fact one leading school of thought argued that designers should not even be required to consider the means by which their creations should be produced. It was claimed that production specialists should be sufficiently skilled to find ways to produce and that design should not be handicapped by such conditions.

Whilst this view currently finds less support than previously, there can be no doubt that the links between design and production are weaker than they should be. It should be obvious that if a design is difficult to produce, it is certain to be more expensive than necessary, and quality problems are likely to be greater than an alternative where the means to produce have been taken into account. Otherwise we may find that we have designed a perfect product which is perfectly impossible, or unnecessarily difficult to produce. Again the Japanese have some answers. There, process design is an integral part of product design. A product is not offered for sale until the process has been thoroughly tested first. In the West it is not unusual to find products being launched or publicised when they are nothing more than a few lines on a piece of paper. Orders are taken even before prototypes have been made let alone tested. This stimulation of the market leads to enormous pressure being brought to bear on the design and developement function to complete the process in the minimum time. This pressure leads to inevitable shortcuts both at the reliability testing stage, and subsequently design for production.

The resulting products, loaded with problems are presented to the market to be tested by the user. Frequently, and particularly with novel or long awaited products, the inevitable failures occur in a blaze of publicity.

More prudent competitors, usually Japanese, do things differently. They know that it is not important to be the first into the market, but it is important to be the first into the market with a good product. This mentality does not lead to spectacular publicity or rumblings in the gossip columns of specialist magazines or journals, but it does lead to a constantly growing reputation for sound development, and confidence.

Accepting the above arguments, it can be seen that a knowledge of variability in production is of great importance to the designer if later problems are to be avoided. The information usually comes in the form of process capability data.

Since at least 50% of component items are usually sub contracted, the need for the acquisition of this information extends beyond the boundaries of the designing organisation, and into the sub contract organisations and sub sub contractors etc. This again raises questions regarding the means by which supplier quality assurance is carried out.

Supplier quality assurance based upon Allied Quality Assurance Publications (AQAP's) and soon to be the basis of I.S.O. 9000, does not require extensive study of processes or their capabilities. The evaluation is mainly concerned with management systems, documentation and inspection procedures.

Ford Motor Company have recognised the weakness of this approach, and, based upon studies in Japan, particularly the Toyota Motor Company, they now weight 60% of their assessment on process capabilities.

Variation Control Methods

Variation is classified in two catagories

1. Variable data.

2. Attribute or Countable data.

Variable Data. This data included any measure which varies on a continuous basis, e.g. temperature, speed, noise level, length, weight, voltage, current, etc.

Attribute Data. This included all data which can be identified as good/bad, right/wrong is/is not etc. It also includes SUBJECTIVE DATA i.e. data which is subjective to the human senses such as smell, feel, appearance, etc.

Some data can be presented in either form. For example length may be measured precisely or with a limit gauge which simply indicates whether or not the dimension is within acceptable limits but does not indicate the precise dimension.

Given that we have a choice, should we measure as variable or attribute? As with most things, there are trade offs.

Usually, attribute data collection is quicker and requires less skill to determine than in its variable form. However, it provides less information, and as a general rule, we need about ten times as much data in attribute form to acquire the same knowledge of the process that we could have gained from variable data.

For Example:

Suppose an operation is required to produce components 10mm in length $^+/-$ ·01mm. and we check them both with a limit gauge and with a micrometer, the results of a sample of 5 items from the batch shown in Table 1 below.

We can see that the limit gauge only told us that the items were good. The micrometer showed that although the items were good, the scatter of the sample of 5 ranged from one limit to the other.

Clearly it is highly likely that there are others in the batch which are outside the limits. This could not have been assumed from the attribute form of the data.

In Western Society, the collection and analysis of variable data for process improvement is far less in evidence than in Japan. Even when it occurs, the data is usually collected by technical specialists and the interpretation not understood by the operating forces. In contrast in Japan, not only is data collected, analysed and used on a massive scale, the majority of the data is handled by the operating forces. Furthermore, even the least skilled operator is likely to be able to interpret standard deviation and other parameters related to the distribution of data.

From a glance at the programme for this course, it is obvious that this paper should not concern itself with the mathematics relating to the normal or other distributions. It is assumed that these are familiar to the reader, or can be easily obtained from standard texts (see references).

It is felt that better use can be made of the participants' time, if the author shares his experiences not in a detailed explanation of the techniques involved in meaning variability, but in the means by which these techniques can be presented to, and used by the operating forces.

TABLE 1

Micrometer	Limit Gauge
10.01	good
10.009	good
9.99	good
10.00	good
10.01	good

PROCESS CAPABILITY

In the context of Production Quality Control, the first application of data collection and analysis is usually for the purpose of establishing process capabilities. Clearly, it is important to establish the degree of inherent variability in processes, both for planning purposes, and as a basis for control. Ideally control should be carried out by the operating forces. To achieve this it is necessary for these forces to have an understanding of the basis of data. However, the majority of people at this level have an aversion to the learning processes usually employed to convey the concepts involved. The standard deviation for example, is a simple concept for those with a mathematical ability but difficult for those not academically inclined. This does not present an insurmountable problem however, provided the matter is approached in the right way and a number of basic rules or assumptions are accepted.

Most people are capable of learning quite complex ideas providing:

a) They want to learn.

b) The material is presented in a manner which is acceptable to the recipient.

c) They can see how they themselves will benefit from the skills acquired. This benefit does not imply direct financial reward, but does imply that work will be more interesting and more rewarding with the acquired skills.

Taking each item:

a) It is assumed that most people want to learn, but have been put off the process by previous bad experiences in the learning environment. It is accepted that this assumption may not be true for every single individual.

b) Non academically inclined people usually have an aversion to numbers and calculations. Experience shows that almost all the features of Process Capability can be presented to and used by even the most non academic without requiring a single calculation. But even the calculations and use of statistical tables can be presented and understood eventually provided that the correct steps are followed. The author has made very significant breakthroughs by using the following approach.

1) Three day training course for line personnel on site to the following schedule.

- Explain difference between attribute and variable data.

- Use course participants' experience to identify examples of each in their own work environmment.

- Use sampling aids to enable participants to produce frequency diagrams for both attributes and variables (see Table 2 on following page).

Explain the reason why the two shapes are different.

- Send participants into their departments in teams of two for the collection of real data in this form.

- Each team to present their results and interpretations to other participants (see Figure 1 on following page).

254

TABLE 2. Example of Attribute data - Scratched Cabinets

No per shift.

0	//// //// //// //// //// //// //// ////
1	//// //// ///// //// ////
2	//// ////
3	////
4	
5	

or

0	////
1	//// ////
2	//// //// /
3	//// //// ////
4	//// //// //// //// ////
5	//// //// //// //// ////
6	//// //// ////
7	//// //
8	////
9	////

FIGURE 1.

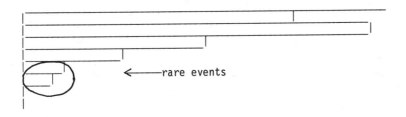

←——rare events

255

- Show how the "tails" of these results can be used for control purpose i.e. observation of known rare events.

- Show how tables can be used to identify probabilities if there were events, but only give superficial explanation of Binomial and Poisson data. No attempt should be made to deal with this using a mathematical approach.

- Application to variable data.

- Repeat the previous experiments but this time using variable data. Charts will be constructed, and this will require an explanation of data grouping.

- Participants again collect data from real life situations and present in frequency distribution form or tally check method. Ensure that approximately 100 observations are made, and that they are consecutive e.g.

TABLE 3

6.0	-	6.01	x
6.011	-	6.02	xx
6.021	-	6.03	xxx
6.031	-	6.04	xxxxx
6.041	-	6.05	xxxx
6.051	-	6.06	xxxx
6.051	-	6.07	xx
6.061	-	6.08	x

- Explain the concept of the normal distribution in general terms.

- Discuss the observation that this type of data has two tails. That these can be used for control purposes as before.

FIGURE 2.

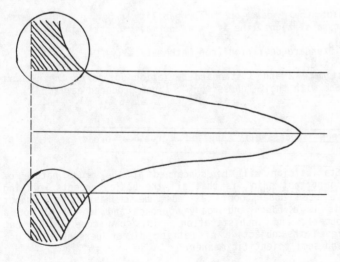

o Participants to estimate the line which separates the tail from the body of the distribution by eye.

  o Introduction to charting. Using previously collected data, participants represent in time to time trend format. Again draw estimated lines or chart.

FIGURE 3

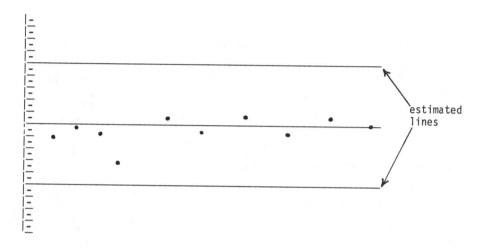

estimated lines

o Explain reasoning behind use of the lines as warning limits.

o Estimate where "Action" limit lines would appear.

o Introduce the concept of probability paper and cumulative frequencies.

o Participants to plot their data on to probability paper. Show how the data approximates to a straight line if it is normally distributed.

o Explain concept of Standard deviation (non mathematically).

o Participants to estimate 2 and 3 standard deviation points on their chart paper. Compare these with their "guess work" lines produced earlier.

END OF COURSE

This approach has been successfully applied across a broad spectrum of industry.

Whilst the trained Statistician will be concerned at the crude nature of the end product, the important point is that all the basic concepts relating to sampling will have been understood. It must be emphasised throughout the training that this is a rough and ready approach and is capable of a great deal of development, and sophistication. This can come later, but at least it has achieved the objective of getting direct personnel to use data in an objective and semi scientific manner.

USE OF PROCESS CAPABILITY DATA

In the West the traditional use of process capability data has been as a basis for control, and for determining the relative precision of processes related to specification tolerances. The formula is:

$$\frac{T}{6\sigma}$$ where T = Specification tolerance.
σ = Standard deviation.

If $T \doteq 6\sigma$ The process is said have <u>medium</u> relative precision.

If $T < 6\sigma$ The process is <u>low</u> relative precision.

If $T > 6\sigma$ The process has <u>high</u> relative precision.

Obviously high relative precision is desirable but it is frequently argued that such an effect results in underutilisation of a high precision, or expensive resource.

Medium precision, whilst appearing to be adequate, has the disadvantage that very precise control of the mean or dispersion is required to avoid out of tolerance conditions. These controls would need to be very much more sophisticated than those covered in the three day course described earlier, and the possible drift of the process must be considered.

Low relative precision is obviously undesirable, and such a process should be avoided where possible. Use of such a process would require rigid control during operation, and the resulting product would require sorting afterwards. Generally it is regarded safe if the process capability is 0.75 of the tolerance.

PROCESS CAPABILITY DATA IN DESIGN

In recent years, the Japanese have adopted the widespread practice of describing the process capability in terms of the ratio of the number of standard deviations in variation to the number of standard deviations in the width of the tolerance band.

FIGURE 4

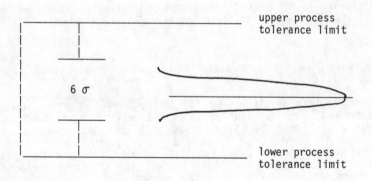

upper process
tolerance limit

6σ

lower process
tolerance limit

This is known as the capability index.

Both internal processes and those of suppliers are judged by this ratio. The higher the figure, the better the process. Japanese suppliers use process capability as an aid to selling and aim to achieve values above 1.33.

The ultimate aim is to eliminate variation altogether. The Japanese maintain that this is the only way possible to achieve parts per million defect levels (ppm's). It is also an essential feature of "Just In Time Production" or "Zero Inventory Concepts".

Competition between suppliers includes the achievement of more impressive capability indices.

This data is used by the design function as part of process selection, and as part of life cycle costing including the Taguchi method currently being evaluated in the West.

QUALITY IMPROVEMENT IN PRODUCTION

Quality in production is determined by three factors (ref Dr Juran).

       1.   Quality Planning
       2.   Quality Improvement
       3.   Quality Control

The overall level of quality achieved is determined by the effectiveness of <u>Quality Planning</u>. This can be shown graphically as follows:

FIGURE 5. Feedback of data for future planning.

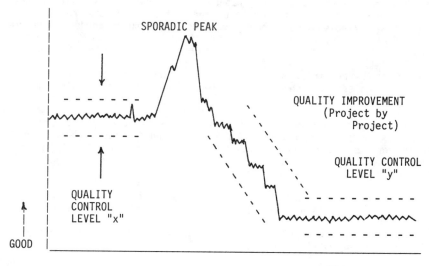

This planning is carried out prior to operations and is based upon information known at that time. Inevitably, of course, many deficiencies will exist which collectively determine the actual level achieved. This is level 'x' on our diagram.

When production takes place, control is then applied to ensure that product quality is held at that level. Occasionally, an adverse change may take place which brings about a loss of control, such as happened recently at the Chernobyl power station. This loss of control brings many forces to bear on the out-of-control situation, in order to restore normality. This activity is usually known as "fire fighting" or "trouble shooting", and relates to sporadic problems.

However, whilst planning, control and firefighting are adquate to hold current and past levels of quality, they are inadequate to meet the needs of the future. As we progress into the future, one thing is more or less certain. The demand for ever improving levels of quality, across the entire product spectrum will increase and at an ever increasing rate. This improvement will happen anyway but at an evolutionary rate; however this will not be rapid enough for survival in the future. During the past decade the world has swung dramatically, from being a suppliers' market to a consumers' market. The main reason is the sudden upsurge of productive output from the Far East nd particularly Japan. This has created upsurge over capacity in most product markets.

In a sellers' market, when we can sell everything we can make, quality may seem relatively unimportant. In a buyers' market, quality assumes equal status to the financial forces, since both will equally determine survival. The lesson is that we must take revolutionary steps to make quality improvements. The evolutionary approach is no longer good enough to ensure survival.

Revolutionary improvement does not happen by chance. It requires both organisation and a better understanding than hitherto of the relationship between process variables and product results. In other words "a voyage of discovery", and it is very different from day to day firefighting activities.

With firefighting, the need is to locate the changes which occurred prior to the sporadic bad effect, and then to determine which of them was the primary cause.

In the case of quality improvement, it is necessary first of all to identify the theories as to the relevant elements of the chronic effect, to test those theories through experimentation, and then to develop, test and implement the remedies. Following this, it is necessary to hold the gains by introduction of control at the new level.

Many chronic problems exist not because the problems are necessarily difficult to solve, but because it is no-one's responsibility to solve such problems.

For example. Consider a typical situation in a manufacturing company at the regular monthly production meeting.

The manager of a production line is being asked to explain why his production is constantly below the agreed target level.

One defensive theory offered by the manager is the suggestion that machine maintenance is inadequate, and it is the responsibility of the Maintenance Department to concentrate more effort in this area.

The Maintenance Manager offers two alternative defensive theories of his own. Firstly, he accuses the line manager of failing to train his people to use the equipment properly, and also for poor housekeeping. Secondly, he may suggest that the Personnel Department refuse to allow him sufficient fitters, electricians, etc, to do the job properly.

The Personnel Department may then accuse the Finance Department, and so it goes on. The problem remains in dead centre because it is no-one's job to solve the problem, and it probably cannot be solved by debate. First of all we must find the facts. It is also necessary to move from an atmosphere of blame to one of collaboration.

Ideally, such problems are best tackled by project teams comprising everyone who has a vested interest in the problem. In the previous example, all of those mentioned should be in the team. Each has his own theories, and these need to be identified and tested. This requires organisation, facilities, resources and time.

Sometimes the theories can be tested by the team members themselves, frequently however, the need for data collection and experimentation goes beyond the time they have available. In this case, it is necessary to bring in full time diagnosticians to provide this service. One of the advantages of using full time personnel relates to the question of objectivity. Such personnel have no vested interest in any of the theories and can therefore adopt an objective approach to the results of data collection and analysis.

Eventually, the cause of the problem is discovered. In the case of the problem described above, let us suppose that the principle culprit was found to be a particular type of electric motor used which is subject to frequent breakdown. Once such a cause is known, the responsibility for solving that problem becomes clear. The engineers are specially trained to solve such problems. It would be their job to improve the design or seek a more reliable alternative. This would not be the responsibility of the project team, but the appropriate department concerned. However, the team should be responsible for ensuring that the remedy proved to be effective, in other words verification.

Following the implementation of the remedy, the project team should also be responsible for designing and implementing the appropriate controls for holding the gains from the improvement and for long term monitoring.

Quality Improvement such as that described above, should not be regarded as a "one off" type activity. Being competitive in the future requires that quality improvement takes place project by project on a continuous basis. It must become a way of managerial life for managers both collectively and individually. The tools required are numerous and several are mentioned in the following text. The 'core' techniques ("brainstorming" "pareto analysis" and "cause and effect analysis") are discussed briefly under the section on Quality Circles which appears at the end of this paper. Others include data collection and analysis, design of experiments and techniques, and other forms of statistical analysis. (Further reading: Managerial Breakthrough - Dr Juran).

QUALITY CONTROL

Whether the process of Quality Improvement is used or not, some form of quality is necessary in order to identify likely sporadic problems as soon as they arise, and to detect adverse trends in the process. The subject of control is quite large and a detailed treatment goes well beyond the scope of this paper. Therefore, we can only deal with generalities or basic principles.

THE CLOSED FEEDBACK LOOP

Control is achieved through the closed feedback loop principle. This is true whether we are talking about a servo mechanism, a process or even people.

The control loop is as follows:

1. <u>PLAN</u> - Determine the target level of quality and decide the acceptable variation from the target level.

2. <u>MEASURE</u> - Take measures to determine what is the actual performance.

3. <u>CORRECTIVE ACTION</u> - This is the effector which is energised when actual performance differs from planned performance by a significant amount.

FIGURE 6. The Loop

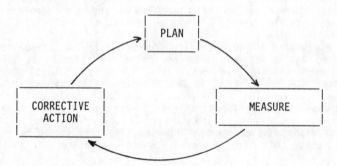

Let us look at each in turn.

1. <u>PLAN</u>

In the case of statistical process control, the <u>planned level</u> is usually the mean or average size required for a given feature or sum of features, and the acceptable variation either side or to one side of that desired value. This is true both for attribute or variable data. In the case of attribute data, we are usually concerned with its use in terms of "number of defects", faults or errors. In most cases the desired level is zero, but since this is not always practicable, some target figure greater than zero must be agreed, and this becomes the planned level.

For variables, which include such features as voltage, current, resistance, speed, temperature, dimensions, weight, etc. the most common situation is the need to hold as closely as possible to some target value. Since some variation is almost bound to occur, it is necessary to agree on the admissible variation either side of that mean. In this case, the mean and the variation become the parameters of the plan.

2. MEASUREMENT

The subject of measurement is extremely broad, since each individual situation will require its own unique form of measurement. However, some principles are worth consideration.

First of all, any form of measurement has its own built-in variation which will add to the variation inherent in the process being monitored. If the measurement is taken by a person rather than automatically, then person-to-person differences must also be considered.

Maintenance of measuring equipment is also a potential source of trouble since many may exhibit a drift from the original values over a period of time. This may be due either to characteristics inherent in the equipment itself as a result of its design, for example the readings given on an electronic comparator may vary over time due to internal heating; or, in the case of mechanical equipment, through wear, vibration or some other cause.

A further consideration relating to variation of measure is the question of traceability. Many expensive quality-related problems have resulted from "gauge to gauge" differences. This can be avoided by ensuring that all gauges are properly calibrated and checked against a known standard. Ultimately, such standards themselves should be checked against national or natural standards. This chain of calibration is referred to as TRACEABILITY.

Also consideration should be given to the point at which the measurements are taken. Sometimes, these are taken at the point at which the operation occurs, and the results are immediately known to the process operator assigned to control the process. In other cases, the measurements may be carried out in a laboratory well away from the point of manufacture. In such cases it is important to get the information back to the operators as quickly as possible and in a form they can understand.

When the information is required for control purposes this becomes even more important.

3. CORRECTIVE ACTION

As with measurement, the question of who or what takes the corrective action is extremely broad; and therefore we can only deal with basic principles in the paper.

For the control of most processes, we are dealing with two basic forms of variation:

- chance causes
- assignable causes.

Chance causes follow the laws of probability and are inherent in every process. Superimposed upon these chance causes, are the variations assignable to some specific feature or features. Any variation, from the mean or target value, when it occurs, could have arisen as a result of either of these causes.

Since the chance variations are inherent in the design of the process, they cannot be eliminated during operations through the application of process control. This is only possible for assignable causes.

It follows, therefore, that before satisfactory process control becomes possible, we must have some means to distinguish between chance and assignable causes.

To do this, we must first of all identify the characteristics typical of both chance and assignable variations in order that they may be identified as they occur. Let us consider both.

a) Chance Causes

Chance variations usually occur with well defined distributions. For example the Binomial distribution and Poisson distributions are the most important when dealing with attribute or Camtable data, and the normal or Gaussian distribution when dealing with variables. A full treatment of these would be out of context with the rest if this paper, but references are given for further reading.

b) Assignable Causes

The three most common forms of assignable cause are:

i) Cyclic Variations. These are often caused by some periodically recurring event such as a stretched chain, refilling of a hopper, thermostat, influence of an adjacent process, etc.

ii) Step Variations. Typical causes are chipped tool, sudden surge of current, oversized components, jumping of a tooth, etc.

iii) Drift. This is common to a great many processes and is usually attributable to wear, overheating effect, etc.

The most popular technique developed for separating the chance causes from the assignable causes, and for enabling corrective action to be taken on the assignable causes is the SHEWHART CONTROL CHART.

This was developed by Dr Walter Shewhart prior to World War Two, and is now widely used throughout the world.

The SHEWHART CONTROL CHART is designed to enable the separation of chance variation from assignable causes during the process as these variations occur.

Control is effected by the use of a chart constructed to show the time-to-time variation of samples averages, and a further chart to plot the range of within sample variation as shown in Figure 7.

Control lines are plotted on the chart based on calculations derived from a PROCESS CAPABILITY STUDY. These control lines are intended to indicate whether a point outside the line is likely to have occurred by chance, or is evidence of an assignable cause.

The two minor lines, known as Warning lines are usually placed at two standard deviations from the mean, such that only one in 40 plots on average will fall on or beyond the line by chance.

FIGURE 7.   Shewhart Chart

SAMPLE MEAN

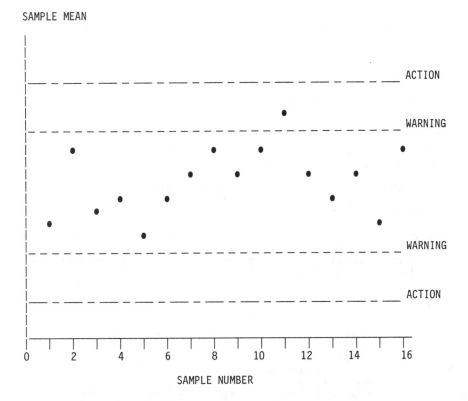

SAMPLE NUMBER

Two further lines known as <u>Action Lines</u> are placed usually at 3 standard deviations from the mean. Since only approximately one in one thousand plots will fall beyond these lines by chance, it is concluded that any such plots will indicate a change in the process.

QUALITY IMPROVEMENT IN PRODUCTION

Control for variables is usually based upon the x.R. Chart or Shewart Chart. Occasionally the CUSUM Chart may be used; but this tends to appear in the more specialised situations where the detection of small changes is critical. The CUSUM technique in its simplest form is extremely easy to use and has much wider application than the current level of use may indicate.

Control for attributes may vary between 'n' 'p' and 'c' charting methods depending upon whether the data is POISSON or BINOMIALLY distributed. Again the CUSUM technique may be used but is less in evidence.

The use of Control Charts represents a major difference in approach to the control of product quality between the West and Japan. This difference appears in two ways.

1). There is little evidence of the use of these techniques generally

265

in the West other than in the current approach being driven by the Ford Motor Company through its suppliers.

In contrast, virtually all Japanese factories use these techniques in every conceivable operation and the theories are well understood at all levels including the operating forces.

2) Even where the concepts are used, they are used for different purposes in Japan from the West.

In the West, the Control Chart concept is used mainly to hold a process within the specification requirements. If it happens that the process capability is high compared with the specification limits, it is highly likely that modified action and warning limits will be applied to give more latitude for variability, or to put it another way, to avoid unnecessarily tight control.

The Japanese would not do this. They use the Control Chart mainly as a diagnostic tool in order to reduce variability regardless of the specification limits.

Any data which shows adverse trends, drift or points outside the warning or action limits would become targets in quality improvement projects, however infrequently they may occur. The ultimate goal is the total elimination of variability.

SAMPLING AND VENDOR CONTROL

The options usually considered for the control of incoming goods include:

1) Supplier evaluation and approval - Quality systems
2) Pre-delivery inspection and test.
3) Incoming inspection and test.

OPTION 1 - SUPPLIER EVALUATION AND APPROVAL

Whilst there are other approaches such as purchase of certificated products approved to trade or National Standards (for example the B.S. Kitemark), the tendency in the West has been to increasingly use Option 1, the approval of a supplier's or potential supplier's quality system.

This approach is not without its drawbacks however. The worst of these are as follows.

1) Variation in assessment results from variability in the skills and perceptions of assessment personnel. This in turn leads to a lack of confidence in the approach.

2) The purchasing or commercial involvement in these activities tend to create a contractual or adversary relationship between the supplier and the customer.

3) There is much evidence to suggest that if the underlying attitudes in the supplier organisation are not quality-orientated there is a tendency to create two systems one that is shown to the assessor, and the other which represents the way things really happen.

Again, there is a real difference between the way the Japanese approach this problem, and the way it is approached in the West.

In the West, there is a tendency for large Companies to "multiple source." In other words, to spread their requirements amongst a wide range of suppliers, to control their input by threats, and source switching as a means of punishing those who fall short of requirements. This is usually effected through vendor rating schemes.

In Japan, there is a growing tendency is to concentrate supplies through a small number of suppliers, who are offered long term contracts. In return, these suppliers are expected to participate in collaborative activities aimed at the overall and continuous improvements of products and processes for their mutual benefit.

Because the number of suppliers is small, it becomes easier to offer technical and other support to the supplier to ensure achievement of goals.

OPTION 2

This option usually forms part of Option 1 and is applied in certain contracts where the product cna be readily assessed at final inspection. Alternatively, some complex electronic equipment which may form part of a large expensive installation may require some form of acceptance testing prior to shipment. This is to avoid extremely costly and time consuming work away from the site of manufacture if faults are subsequently found to be present. This is typical of nuclear industry equipment.

OPTION 3

This appraisal is rapidly being replaced by more sophisticated supplier evaluation programmmes for several reasons.

1) It is being widely accepted that the responsibility for conformance should rest with the supplier not the customer.

2) The demand for defect levels in the order of parts per million (ppm) rather than percentage or parts per hundred means that sampling becomes a non-solution.

    For detection of faults at ppm levels, the sample sizes become equal to the lot sizes, and so sampling becomes academic. Even then, the fallibility of the human inspector is such that non-human forms of automatic inspection must be used to detect such defects.

WARRANTY AND LIABILITY COSTS

Society has become intolerant of defects no matter how rare. This is particularly true where human safety and health are involved. Additionally, product liability laws in most industrialised countries are now moving towards strict liability. In other words, "If the product causes injury, it is regarded as defective even though no particular fault is present."

An organisation can adopt several options to reduce the risks:

- Avoid it
- Reduce it
- Pass the risk to others

The risk is avoided if the producing organisation ceases to produce the offending type of product. For example, the risk of being prosecuted for the effects of blue asbestos can be eliminated by the use of an alternative material.

We can reduce the risk by operating a system of effective Quality Assurance at all stages in our organisation.

The risk can be passed to others through insurance. However, there is an increasing tendency for insurance companies to relate the premiums to the risk involved. These are affected both by the type of product, and the level of quality assurance.

In the future it is highly likely that insurance companies will demand that their customers or clients should operate accepted quality assurance systems before insurance will even be offered.

QUALITY CIRCLES

Quality Circles or direct employee small group activities are growing rapidly in the West.

The Principle of 'Self-Control'.

Most Western people confuse Quality Circles with problem-solving groups. Quality Circles in fact represent a form of 'self-control.' This ultimately goes far beyond simple problem solving, although problem solving is usually the point where they start.

The size of the group that comprises a Quality Circle is important. Too large a group makes it difficult for everyone to participate. If there are more people in the work area than can be accommodated in one Circle, there is no reason why others should not be formed to involve the remaining employees. They are unlikely to conflict with each other. Most Circles have between 6 and 10 people, and the concept works well with groups of this size. Sometimes there are Circles of only 3 to 5 people simply because there are only that number of people in the section.

CIRCLE DEVELOPMENT

Once a Circle has been formed, all being well, it will pass through three distinct phases of development, to the fourth ultimate stage. Whether or not it ever reaches this final stage is entirely dependent upon the objectives and support of management.

The problem-solving phase

Phase 1 - Initial phase. During this phase, the Circle will have been trained in simple techniques which will enable its members to identify, analyse and solve some of the more pressing problems in their own work area. These problems will include:

> Wastage of materials;
> Housekeeping problems;
> Delays, hold-ups, etc;
> Inadequate job instructions;
> Quality;

Productivity;
Energy consumption;
Environmental problems;
Handling;
Safety; and
the quality of work life generally.

These will usually be the problems that are uppermost in the minds of most employees.

Phase 2 - Monitoring and problem solving. After a short time, when several of the simpler problems have been resolved and many others have just 'disappeared' as a result of other improvements in the work environment, the Circle will begin to develop a 'monitoring' mentality. The members will have been trained in simple control techniques, and will use these to maintain the improvements already made.

Phase 3 - Innovation - self-improvement and problem solving. There is almost a natural progression to the self-improvement phase from Phase 2. As the Circle begins to mature, and most of the techniques taught have been well practised and understood, the confidence of the group will have grown considerably. The members will also have gained a wider acceptance by their colleagues in their own and other departments and also by management. In other words, they will be treated with greater respect.

Phase 4 - Self-control. At the time of writing, it is unlikely that any Circles in Western Europe have reached this stage of development. Whether it is reached at all is as much dependent on managers and others outside the Circle as it is on the Circle members themselves.

Having passed through Phases 1, 2 and 3, the Circle will be very mature, and trusted by management. The organisation will have realised much of the potential available from this style of management, and will be both seeking ways of furthering the development of the existing Circles and envisaging new ones.

The latter is only a question of continuing the same form of development with new groups, but the continuous development of existing, mature Circles will be breaking fresh ground in most societies, and is relatively recent among even those companies with 20 years of development. The development of existing Circles involves two factors, internal and external.

Internally, it is necessary for the organisation to ensure that such Circles have access to all the information, training aids and techniques necessary for them to progress. They may indulge in self-study. It will be necessary for management consciously to give them information such as Quality Control data and so forth. They should have access to technical journals relating to their work and attend in-house seminars in order to be kept abreast of the latest developments in their field.

Externally, such Circles should be given the opportunity of commmunicating with professional educational and specialist institutions, and should make either direct or indirect contact with suppliers when relevant to their activities. They should be permitted to attend conventions where they can meet Circle members from other organisations, and can trade experiences with each other. They should also be allowed to attend conferences and seminars to help them progress in their work.

Let us now consider the definition of a Quality Circle:

"A Quality Circle is a small group of between three and twelve people who do the same or similar work, voluntarily meeting together regularly for about one hour per week in paid time, usually under the leadership of their own supervisor, and trained to identify, analyse and solve some of the problems in their work, presenting solutions to management and, where possible, implementing solutions themselves."

Taking each aspect of the definition in turn, the basis of each point will be reviewed.

1. 'A small group of people who do similar work'

The Circle should comprise a more or less homogeneous group of people, usually from the same work area. They will usually have a similar educational background, speak the same work language, and no one member should be inhibited in any way by the presence of another. Whilst in exceptional circumstances some variation of this important rule may be necessary, generally speaking, such variations should be avoided.

Circles comprising members from different disciplines or with different work experience will find it difficult to select a project which is of interest to all members. Those least directly affected by the potential achievement will show less interest than the other members, and, if some of the work-related jargon is unfamiliar, boredom may be induced, with a resulting loss of morale and possibly the loss of some members.

That is not to say that the Circle cannot utilise the services of specialists or others if it so wishes. For example, the members may invite such people into the Circle for a specific project if they feel it will help produce a more soundly thought out solution. In such cases, these guests are really acting as consultants to the Circle, and a supportive management should actively encourage this process.

This arrangement fits in well with the concept of 'self-control' which should be the ultimate aim of corporate management in developing Circles in the first place.

2. 'Three to twelve people'

If the work area contains more enthusiasts than can be included in one Circle, others may be formed progressively once the earlier groups have become established. Those as yet unfamiliar with Quality Circles specifically may fear that such a development can lead to conflict and rivalry between groups, but this is extremely rare. It is far more likely that they will co-operate with each other, even helping to collect data for each other's projects, and occasionally, if need arises, form cross Circle sub-groups for the solution of specific problems. Such developments are a sign of maturity in Circle activities and are to be encouraged wherever possible.

In cases where there are only one, two or three people in the work area, it may not be possible to form them into a Circle, but usually, they will have considerable interaction with other more heavily populated sections.

3. 'Voluntarily meeting together'

The meaning of the word 'voluntarily' is hard to define, but basically, in the context of Quality Circles, it means that no one has to join a Circle. People are free to join and free to leave. If someone joins a Circle and subsequently chooses to leave it, there should be no pressure, inquests or recriminations. Obviously, if someone drops out of a Circle it should be regarded as a danger signal that all might not be well in the group, and the Circle leaver should be discreetly asked the reason for leaving. If there is a problem, and it can be overcome, then that individual may, if he chooses to do so, return to the group if the opportunity exists.

Whilst the number of volunteers may be quite small in the early stages, when people may possibly be suspicious of management motives, the number should begin to increase dramatically as soon as the achievements of the earlier Circles become known and confidence is gained.

The fact that membership is voluntary does not mean that the organisation has to wait until people knock on the door and request a Circle to be formed. Many people, possibly most, in the early stages have been invited to join but not compelled.

They should be free to drop out at any time if they wish, even in the middle of training. In a sense, they are actually only volunteering to attend the next meeting, although dropping out is fortunately relatively rare.

4. 'Meeting regularly for about an hour per week'

Whilst some variation in timing exists, it is generally agreed that when circumstances permit, the regular weekly meeting is preferred to once fortnightly or to irregular times on a weekly basis.

A regular meeting time is habit-forming, and the day of the meeting will soon be associated in the minds of the members as 'Circle Day' and in such cases, members are much less likely to forget to attend. This sometimes happens in other cases.

5. 'In paid time'

This may be particularly relevant in shift work operations, when Circles may sometimes span shifts. If the Circle comprises members from each of three or more shifts, it may be possible to hold the meeting during an overlap between two shifts, but the members from other shifts will either miss the meeting, or have to attend outside shift time.

6. 'Under the leadership of their own supervisor'

Management's motives in setting up Quality Circles are to make better use of the existing structure, not to create alternatives. Circles are concerned with work-related problems and not with grievances, wages, salaries or conditions of employment. If these items are contentious then the group must take them up through the appropriate channels in the usual way. Circles are not part of the bargaining, negotiating or grievance machinery, neither do they impinge upon the activities of those who are responsible for these aspects of a company's affairs.

Because the Circle is purely concerned with work-related problems, and because the supervisor is the appointed leader of the group, it follows that direct supervisors should at least have the first option to be the Circle leader.

However, once the group has been formed, the members, and others in the work area will quickly realise that is doesn't matter who the leader is because Circle decision-making is a totally democratic process. When the Circle members are in the meeting room together, everyone has one vote, and no one's opinion is any more or less important than anyone else's.

7. 'To identify, analyse and solve problems in their work'

The key point about this part of the definition is the fact that the Circles identify their own problems in their own work area. That is not to say that other people may not make suggestions. Indeed, the essence of Quality Circles ultimately should be that the Circles really become managers at their own level.

It is this aspect of Quality Circle activities which gives the members the greatest satisfaction. Because they are not meeting to criticise the work of others, they find that they can make real progress with their projects. When asked what they like most about Quality Circles, one of the most frequent answers comes back: 'we find we can get things done'. 'These problems have been around for years, and now we are making progress.'

8. 'Presenting solutions to management'

The group is usually proud of its achievement and the teamwork involved. It will probably have worked very hard, may have spent lunchtimes, evenings or even week-ends working on its ideas if its members have been enthusiastic enough, and frequently they are. Consequently, the presentations of their ideas to management are the culmination of all this activity.

It would be unfortunate if they were unable to convince their manager of the benefits of their ideas, simply because they were badly presented or because the members were forced to present their ideas in the form of a report which might not be read. Therefore, training newly formed Circles in presentation techniques is extremely important. They may use two or even three meetings to plan and prepare their presentation.

9. 'Implementing the solutions themselves'

Because Circles are usually concerned with problems in their own work rather than with those over the fence in the next department, they can often implement the solutions themselves. This is particularly true of housekeeping problems, reduction in waste material, energy saving and so forth. They also frequently find better ways of doing their own jobs. For example, a Circle in the credit control department of a division of a fairly large company formed a Quality Circle. For their first project the members decided to analyse one of their work routines that they found to be particularly tedious. The result was that they reduced the work content by 16 hours per month. In the process of this work they highlighted another problem which when solved, saved a further 17 hours a month, making a total monthly saving of 33 hours. For their third project, they decided to brainstorm all the possible ways they could make use of the time saved. Someone suggested that they might follow up the invoices with a telephone call. The effect of this idea was to reduce the average credit period by nearly two weeks, thereby making available to that company a considerable sum of money.

BASIC QUALITY CIRCLE TECHNIQUES

It is important not to give Quality Circles any training which cannot be immediately assimilated and used. Circles have floundered more frequently as a result of over ambitious training programmes, than by the reverse. The key phrase is 'keep it simple', and this also applies to the entire programme. There is a tendency for the more sophisticated consultant to overload the training programme with discussions of the relative merits of Hertsberg, Maslow, McGregor, etc, but I think that such tendencies should be vigorously resisted. Quality Circles are basically a simple concept, and should remain so. In fact, simplicity is the main attraction.

It is also important that the person who conducts the training should be able to present ideas in a manner which can be immediately assimilated by the recipient. For this reason, I shall discuss only the basic techniques.

Pareto Analysis

Between the First and Second World Wars, Pareto, an Italian economist, discovered an almost universal relationship between value and quantity.

In a company store, for example, it can usually be found that 20 per cent of the various items are responsible for 80 per cent of the total stock value. This 80/20 relationship holds good for very many situations, and seems to apply to quality problems as well. Many quality failure investigations have shown that around 80 per cent of the failure costs are attributable to 20 per cent of the problems. Quite often, this fact is not realised, consequently, the remaining 80 percent of the problems obtain more consideration than they are worth. The less frequent, but ultimately more costly problems may be ignored altogether.

The following sequence, known as Pareto Analysis may be used to ensure the most cost effective approach to quality problem solving.

Most companies produce some sort of regular scrap report. These usually contain, in percentage form, a ranking of the most frequent causes of defective work which have arisen since the presentation of the previous report.

Sometimes these appear in histogram form, similar to the following: but instead of plotting frequency of occurrence, we can plot per cent contribution to cost.

FIGURE 8. Per cent contribution to cost £

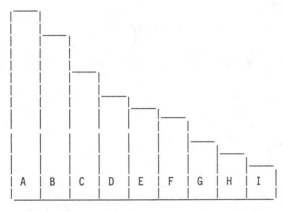

Cause Number

Figure 8 shows that A and B are the most costly items and hence may form the basis of Quality Circle problem selection.

BRAIN STORMING

This is common to almost all Qualtiy Circle activities and Circles around the world make use of the Cause and Effect diagram. This is also variously known as the Ishikawa diagram or Fish Bone diagram.

After the problem has been selected by Pareto Analysis, the Circle may then produce a Cause and Effect diagram. The technique consists of defining an occurrence (effect), and then reducing it to its contributing factors (causes).

FIGURE 9

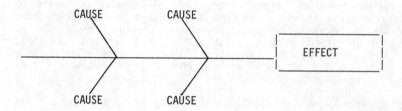

The factors are then critically analysed in the light of their probable contribution to the effect. The factors selected as most likely causes of the effect are then subjected to experimentation to determine the validity of their selection. This method of selection and analysis is repeated until the true causes of the effect are identified.

As with all brainstorming techniques, members of the Group are encouraged to participate equally, and ideas are never ridiculed or attacked in any way. There are essentially four steps in the construction of a C & E diagram.

- identify the effect
- establish the target
- construct the diagram
- contemplate the ideas identified by the diagram.

Step One - Identification may be carried out by either Pareto Analysis, voting, or other means such as Management suggestion or other sources outside the group.

Step Two - realistic targets for achievement are vital to the continued morale of the Circle. Such targets might say 'the elimination of oil spots on plastic film' or, 'a 20 per cent reduction in handling damage before October'. It is important that the group should identify the level of the problem prior to the investigation and elimination of causes.

Step Three - at this point the diagram may be constructed. One very effective way of doing this is to ask each member of the group in turn what he thinks are the causes of the problem. In this way, all members of the group are required to participate. First of all, a horizontal line is drawn with a box which defines the effect. As the members contribute their suggestions

274

FIGURE 10

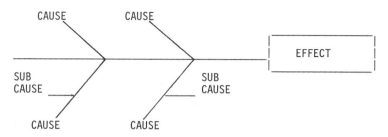

as to the cause, the diagram develops with main causes forming the branch lines, and sub causes, protruding horizontally from the main causes identified (see Figure 10 above).

An easy starting formula to remember is the 4 M's; Manpower, Machinery, Methods, Material. Others may be used when appropriate.

Step Four - After the diagram has been constructed, and this would normally take one or two meetings to produce, it is necessary to allow time for the ideas to clear in people's minds. It is therefore advantageous to wait for a week or a fortnight before investigating the causes, and conducting any experiments necessary to test their validity. Another advantage of this pause is that people will have forgotten who made the various suggestions with the consequence that ideas are attacked rather than people.

Other useful techniques which may be introduced at the initial stages or later if necessary are histograms, scatter diagrams, pre-control etc. Generally, however, considerable care should be taken to ensure that these additional ideas are not introduced too quickly. Many of the recipients will not have attended training sessions since they left school, and probably have little confidence in their ability to learn scientific techniques. It is therefore extremely important to use skills development techniques and to develop confidence before exposure to real problems in a group environment.

FIGURE 11.   The Circle Problem Solving Process

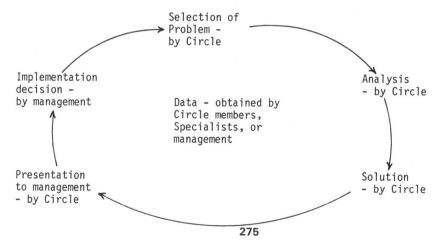

PROBLEM IDENTIFICATION

This can be carried out by anyone in the company and may be the result of customer complaint data, management information, quality control feedback, production engineering or design.

The Circles themselves often identify problems of which Management are sometimes unaware. This is particularly true of handling problems, damage, and shortcomings on the job cards or specification sheets.

PROBLEM SELECTION

It is vitally important that the Groups should be free to select their own problems for solution. The fact that Pareto analysis has been taught as the basis of Circle activities will ensure that the problems selected have a high realistic content.

PROBLEM SOLUTION

This will involve use of the techniques outlined above but often will require consultancy assistance from other groups such as production engineering, quality control and design. It is this greater interest in co-operating with specialist functions which appeals most. In companies where Quality Circles have flourished, production engineers and designers have become enthusiastic because they find the shop floor people more receptive to their own ideas.

RECOMMENDATION TO MANAGEMENT

The manner in which this aspect is conducted will greatly influence the future enthusiasm of Circle Members. Recognition is a powerful motivator, and members will feel ten feet tall when they know they have cracked a problem, and that their achievement has been recognised by management. No one wants to feel that their company makes poor quality products and the pride of achievement in helping to improve the company image, is more important than many people realise.

IMPLEMENTATION DECISION BY MANAGEMENT

It was mentioned earlier that management commitment is essential to the healthy development of Circle activities. It is therefore extremely important that management should give serious consideration the the Circle members' recommendations before rejecting a proposal. Should this be necessary, then the reasons should be clearly explained to Circle members.

WHERE CAN CIRCLES OPERATE?

Quality Circles can be formed anywhere people work together and share similar problems. They need not be restricted to the shop floor. Indeed, in several companies in the United States and in Japan, Quality Circles exist at all levels in the organisation, including clerical workers, and considerable savings have been made by many of these groups.

MEASURES OF SUCCESS

The success of a Quality Circle programme can be measured in three main categories:

Quality - This can be measured by:
Defects/Man hour
Scrap/Unit Manufacture
Customer return data etc.

Cost - The factors here can relate to both quality
and productivity, ie:
Quality Cost Ratios such as:
Failure Cost
Total Cost of Manufacture
Total Cost of Quality
Cost of Sales
Total Cost of Quality

Attitude - Measures of attitude may include improvements to:
Labour Turnover
Absenteeism
Reduction in accidents
Stoppages
Attitude Surveys

RESULTS OF CIRCLE ACTIVITIES

Many of the benefits cannot be discribed in cost terms. Neither is it desirable that they should. Too heavy a concentration on this aspect by Management in the early stages will often distort the spirit behind the programme.

However, cost savings are usually very considerable. Some examples of the annual savings in various companies around the world are:

Savings of time lost due to conflicting job instructions	£165 454
Time lost locating precision tools - saving	£71 363
Elimination of oil leaks contaminating materials - saving	£1 890
Tinplate finish problems - saving	£78 00C
Improvements to grinding process - saving	£102 000
Improved ship-loading methods - saving	£3 236

SOME QUESTIONS ANSWERED

Will Quality Circles be opposed by our Trades Unions?
No. It is very unusual for Trades Unions to oppose Quality Circles. If they are approached properly in the first place most trades unionists see Quality Circles as a way of improving the job satisfaction of their members, and subsequently shop stewards often become very enthusiastic members of Circle groups.

Will Quality Circles only work in large firms using flow production techniques?
No. Quality Circles will work anywhere where people work together and share common problems. This also applies to commercial companies. Banks for example have obtained considerable benefit from this form of motivation.

How often do Quality Circles meet?
About once per week or fortnight depending upon circumstances. The meetings usually last for about one hour.

How do we know that the circles will not cost us a lot of money in time spent talking about problems instead of getting on with the job?
It would be very unusual if this happened. Generally speaking, Quality Circles offer benefit to cost ratios of between 5:1 and 8:1. In addition to this, there are many spin off benefits in the form of reduced labour turnover and increased co-operation between design, production engineering and the shop floor.

We often speak to our shop floor people about quality and production problems, what's so different about Quality Circles?
The fundamental difference is that Quality Circles are actually encouraged to solve problems not just to state them. They also require team work rather than just relying on the isolated aspiration of the occasional confident individual. The team aspect is important. If the group has been properly formulated and the training well conducted, then all the members will contribute their ideas and energy to solving the problems which are often better understood by the man on the job, who is confronted by the problem in his everyday work.

Why should work people who don't normally show any interest in the company's problems suddenly become so concerned?
Usually it is because they have never been asked. We in Britian particularly, have devolved problem solving to management specialists, and have ignored the specialist knowledge which exists on the shop floor. It is this so far untapped resource which gives so much potential to the Quality Circle concept.

Most new management concepts work in the short term but then quickly lose their momentum. What happens once Quality Circles have been operating for a time? Do they suffer in the same way?
The difference between Quality Circles and other aproaches such as zero defects campaigns is that these alternatives do not actually involve people. Nor do they develop the full potential of the individual worker. Quality Circles give every member of an organisation the opportunity to develop his or her own natural talents to their full, and this applies regardless of the individual's specific role in the company.

There is simply no end to the personal development process. By making a foreman or supervisor the leader of a group he will develop his leadership qualities and will grow in stature by imparting training to his group members. As the group develops, his training may be updated progressively and hence increasing both his, and the group's awareness of the complexity of problems, and applying a more scientific approach to their solutions.

It is incredible to realise that foremen in Japanese factories are capable of using degree level statistics to analyse problems. It must be realised of course that Japan has been operating Quality Circles since 1962. It seems therefore that there is virtually no limit to the long term development prospects for this concept. What is perhaps more important is the question, Where will we be if we do not employ the Quality Circle approach? Particularly when it is realised that Circles are now catching on almost everywhere!

We have been offered many panaceas for our problems, how do we know that this isn't just another quack remedy?
The evidence. It is there for all to see. No other management concept has developed at anything like the pace of Quality Circles, and no other concept has so many evangelists at all levels in the company from Managing Director down to the most lowly labourer on the shop floor.

Quality Circles have swept through Japan, and are now sweeping throught Korea,

Taiwan, Brazil, Sweden, Norway and the United States.

Quality Circles are now known to exist in over 52 countries of the world. The concept is more complex than the foregoing description may suggest, and there are many pitfalls.

However, there can be no doubt that those companies which have taken the trouble to introduce the concept properly, are now reaping many rewards.

References

Quality Control Handbook by Dr J M Juran, McGraw-Hill Book Company

Quality Circles Handbook by David Hutchins, Pitman Publishing

The Japanese Approach to Product Quality by Sasaki & Hutchins, Pergamon Press

Quality Planning and Analysis by Dr J M Juran and Dr F Gryna, McGrawHill Book Company

Managerial Breakthrough by Dr J M Juran, McGraw-Hill Book Company

18
■

Maintainability, Maintenance, and Availability

P. D. T. O'CONNOR
British Aerospace
Hertfordshire, UK

INTRODUCTION

Most systems are maintained, ie. they are repaired when they fail, and work is performed on them to keep them operating. The ease with which repairs and other maintenance work can be carried out determines a system's maintainability. Maintainability can be quantified as the mean time to repair (MTTR). The time to repair, however, includes several activities, usually divided into three groups:

1. Preparation time: finding the person for the job, travel, obtaining tools and test equipment, etc.
2. Active maintenance time: actually doing the job.
3. Delay time (logistics time): waiting for spares, etc., once the job has been started.

Active maintenance time includes time for studying repair charts, etc., before the actual repair is started, and time spent in verifying that the repair is satisfactory. It might also include time for post-repair documentation when this must be completed before the equipment can be made available, e.g. on aircraft. Maintainability is often specified as a mean active maintenance time (MAMT), since it is only the active time (excluding documentation) that the designer can influence.

Maintained systems may be subject to corrective and preventive maintenance. Corrective maintenance includes all action to return a system from a failed to an operating or available state. The amount of corrective maintenance is therefore determined by reliability. Corrective maintenance action cannot be planned; it happens when we do not want it to.

Preventive maintenance seeks to retain the system in an operational state by preventing failures from occurring. This can be servicing, such as cleaning and lubrication, or by inspection to find and rectify incipient failures, e.g. by crack detection or calibration. Preventive maintenance affects reliability directly. It is planned and should be performed when we want it to be. Preventive maintenance is measured by the time taken to perform the specified maintenance tasks and their specified frequency.

Maintainability affects availability directly. The time taken to repair failures and to carry out routine preventive maintenance removes the system from

This chapter is extracted from "Practical Reliability Engineering," Chapter 9, by P. D. T. O'Connor, with the permission of the publishers, John Wiley & Sons Ltd.

the available state. There is thus a close relationship between reliability and maintainability, one affecting the other and both affecting availability and costs. In the steady state, ie. after any transient behaviour has settled down and assuming that maintenance actions occur at a constant rate:

$$\text{Availability} = \frac{\text{MTBF}}{\text{MTBF} + \text{MTTR} + \text{mean preventive maintenance time}}$$

The maintainability of a system is clearly governed by the design. The design determines features such as accessibility, ease of test and diagnosis, and requirements for calibration, lubrication and other preventive maintenance actions.

This chapter describes how maintainability can be optimized by design, and how it can be predicted and measured. It also shows how plans for preventive maintenance can be optimized in relation to reliability, to minimize downtime and costs.

MAINTENANCE TIME DISTRIBUTIONS

Maintenance times tend to be lognormally distributed (Figure 1). This has been shown by analysis of data. It also fits our experience and intuition that, for a task or group of tasks, there are occasions when the work is performed rather quickly, but it is relatively unlikely that the work will be done in much less time than usual.

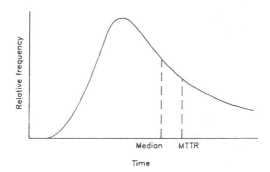

Figure 1.   The lognormal distribution of maintenance time.

In addition to the job-to-job variability, leading typically to a lognormal distribution of repair times, there is also variability due to learning. Depending upon how data are collected, this variability might be included in the job-to-job variability, e.g. if technicians of different experience are being used simultaneously. However, both the mean time and the variance should reduce with experience and training.

(m = scheduled replacement interval)

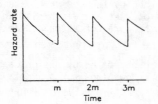

Decreasing hazard rate:
scheduled replacement
increases failure
probability.

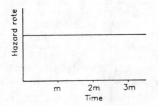

Constant hazard rate:
scheduled replacement
has no effect on failure
probability.

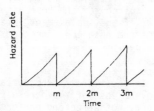

Increasing hazard rate:
scheduled replacement
reduces failure
probability.

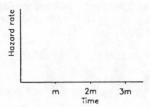

Increasing hazard rate
with failure—free life >m:
scheduled replacement
makes failure
probability = 0.

Figure 2 Theoretical reliability and scheduled replacement relationships (From ref. 1)

Lognormal or Weibull probability paper can be used for plotting maintenance time data.

PREVENTIVE MAINTENANCE STRATEGY

The effectiveness and economy of preventive maintenance can be maximized by taking account of the time-to-failure distributions of the maintained parts and of the failure rate trend of the system.

In general, if a part has a decreasing hazard rate, any replacement will increase the probability of failure. If the hazard is constant, replacement will make no difference to the failure probability. If a part has an increasing hazard rate, then scheduled replacement at any time will in theory improve reliability of the system. However, if the part has a failure-free life (Weibull $\gamma > 0$), then replacement before this time will ensure that failures do not occur. These situations are shown in Figure 2.

These are theoretical considerations. They assume that the replacement action does not introduce any other defects and that the time-to-failure distributions are exactly defined. As explained in Chapter 13, these assumptions must not be made without question. However, it is obviously of prime importance to take account of the time-to-failure distributions of parts in planning a preventive maintenance strategy.

In addition to the effect of replacement on reliability as theoretically determined by considering the time-to-failure distributions of the replaced parts, we must also take account of the effects of the maintenance action on reliability. For example, data might show that a high pressure hydraulic hose has an increasing hazard rate after a failure-free life, in terms of hose leaks. A sensible maintenance policy might therefore be to replace the hose after, say 80 per cent. of the failure-free life. However, if the replacement action increases the probability of hydraulic leaks from the hose end connectors, it might be more economical to replace hoses on failure.

The effects of failures, both in terms of effects on the system and of costs of downtime and repair, must also be considered. In the hydraulic hose example, for instance, a hose leak might be serious if severe loss of fluid results, but end connector leaks might generally be only slight, not affecting performance or safety. A good example of replacement strategy being optimized from the cost point of view is the scheduled replacement of incandescent and fluorescent light units. It is cheaper to replace all units at a scheduled time before an expected proportion will have failed, rather than to replace each unit on failure.

In order to optimize preventive replacement, it is therefore necessary to know the following for each part:

1. The time-to-failure distribution parameters for the main failure modes.
2. The effects of failure.
3. The cost of failure.
4. The cost of scheduled replacement.
5. The likely effect of maintenance on reliability.

We have considered so far parts which do not give any warning of the onset of failure. If incipient failure can be detected, e.g. by inspection, nondestructive testing, etc., we must consider:

6. The rate at which defects propagate to cause failure.
7. The cost of inspection or test.

Note that, from 2, an FMECA is therefore an essential input to maintenance planning.

This systematic approach to maintenance planning, taking account of reliability aspects,is called reliability centred maintenance (RCM).

Example

A flexible cable on a robot assembly line has a time-to-failure distribution which is Weibull, with $\gamma = 150$ h, $\beta = 1.7$ and $\eta = 300$ h. If failure occurs whilst in use the cost of stopping the line and replacing the cable is $5,000. The cost of replacement during scheduled maintenance is $500. If the line runs for 5,000 hours a year and scheduled maintenance takes place every week (100 hours), what would be the annual expected cost of replacement at one-weekly or two-weekly intervals?

With no scheduled replacement the probability of a failure occuring in t hours will be

$$1 - \exp\left[\frac{-(t - 150)}{300}\right]^{1.7}$$

With scheduled replacement after m hours, the scheduled maintenance cost in 5,000 h will be

$$\frac{5000}{m} \times 500 = \frac{2.5 \times 10^6}{m}$$

and the expected failure cost in each scheduled replacement interval will be (assuming not more than one failure in any replacement interval):

$$5,000\left\{1 - \exp\left[\frac{(m - 150)}{300}\right]^{1.7}\right\}$$

Then the total cost per year will be:

$$C = \frac{2.5 \times 10^6}{m} + \frac{5,000 \times 5,000}{m}\left\{1 - \exp\left[\frac{-(m - 150)}{300}\right]^{1.7}\right\}$$

Results are as follows:

	C	No.of scheduled replacements	Expected failures
100	$25,000	50	0
200	$18,304	25	1.2
400	$38,735	12	6.5

Therefore the optimum policy might be to replace the cables at alternate
scheduled maintenance intervals, taking a slight risk of failure.   (Note that
the example assumes that not more than one failure occurs in any scheduled
maintenance interval. If m is only a little more than γ this is a reasonable
assumption.)

A more complete analysis could be performed using Monte Carlo simulation. We
could then take account of more detailed maintenance policies; e.g. it might be
decided that if a cable had been replaced due to failure shortly before a
scheduled maintenance period, it would not be replaced at that time.

Practical Implications

The time-to-failure patterns of components in a system therefore largely
dictate the optimum maintenance policy.   Generally, since electronic
components do not wear out, scheduled tests and replacement do not improve
reliability.   Indeed they are more likely to induce failures (real or
reported).   Electronic equipment should only be subjected to periodic test or
calibration when drifts in parameters or other failures can cause the equipment
to operate outside specification without the user being aware.   Built-in test
and autocalibration can reduce or eliminate the need for periodic test.

FMECA AND FTA MAINTENANCE PLANNING

The FMECA is an important prerequisite to effective maintenance planning and
maintainability analysis.   As shown earlier, the effects of failure modes
(costs, safety implications, detectability) must be considered in determining
scheduled maintenance requirements.   The FMECA is also a very useful input for
preparation of diagnostic procedures and checklists, since the likely causes of
failure symptoms can be traced back using the FMECA results.   When a fault tree
analysis (FTA) has been performed, it can also be used for this purpose.

BUILT-IN TEST (BIT)

Complex electronic systems such as laboratory instruments, avionics and process
control systems now frequently include built-in test (BIT) facilities.   BIT
consists of additional hardware (and often software) which is used for carrying
out functional tests on the system.   BIT might be designed to be activated by
the operator, or it might monitor the system continuously or at set intervals.

BIT can be very effective in increasing system availability and user confidence
in the system.   However, BIT inevitably adds complexity and cost and can
therefore increase the probability of failure.   Additional sensors might be
needed as well as BIT circuitry and displays.   In microprocessor-controlled
systems BIT can be largely implemented in software, but even then additional
memory might be required.

BIT can also adversely affect apparent reliability by falsely indicating that
the system is in a failed condition.   This can be caused by failures within the
BIT, such as failures of sensors, connections, or other components.   BIT should
therefore be kept simple, and limited to monitoring of essential functions which
cannot otherwise be easily monitored.

It is important to optimize the design of BIT in relation to reliability, availability and cost. Sometimes BIT performance is specified (e.g. '90 per cent. of failures must be detected and correctly diagnosed by BIT'). An FMECA can be useful in checking designs against BIT requirements since BIT detection can be assessed against all the important failure modes identified.

MAINTAINABILITY PREDICTION

Maintainability prediction is the estimation of the maintenance workload which will be imposed by scheduled and unscheduled maintenance. The standard method used for this work is US MIL-HDBK-472, which contains four methods for predicting the mean time to repair (MTTR) of a system. Method II is the most frequently used. This is based simply on summing the products of the expected repair times of the individual failure modes and dividing by the sum of the individual failure rates, ie.:

$$MTTR = \frac{\sum (\lambda t_i)}{\sum \lambda}$$

The same approach is used for predicting the mean preventive maintenance time, with replaced by the frequency of occurrence of the preventive maintenance action.

MIL-HDBK-472 describes the methods to be used for predicting individual task times, based upon design considerations such as accessibility, skill levels required, etc. It also describes the procedures for calculating and documenting the analysis, and for selection of maintenance tasks when a sampling basis is to be used (method III), rather than by considering all maintenance activities, which is impracticable on complex systems.

MAINTAINABILITY DEMONSTRATION

The standard approach to maintainability demonstration is MIL-STD-471. The technique is the same as for maintainability prediction using method III of MIL-HDBK-472, except that the individual task times are measured rather than estimated from the design. Selection of task times to be demonstrated might be by agreement or by random selection from a list of maintenance activities.

DESIGN FOR MAINTAINABILITY

It is obviously important that maintained systems are designed so that maintenance tasks are easily performed, and that the skill level required for diagnosis, repair and scheduled maintenance are not too high, considering experience and training of likely maintenance personnel and users. Features such as ease of access and handling, the use of standard tools and equipment rather than specials, and the elimination of the need for delicate adjustment or calibration are desirable in maintained systems. As far as is practicable, the need for scheduled maintenance should be eliminated. Whilst the designer has no control over the performance of maintenance people, he or she can directly affect the inherent maintainability of a system.

Design rules and checklists should include guidance, based on experience of the relevant systems, to aid design for maintainability and to guide design review teams.

Design for maintainability is closely related to design for ease of production. If a product is easy to assemble and test maintenance will usually be easier. Design for testability of electronic circuits is particularly important in this respect, since circuit testability can greatly affect the ease and accuracy of failure diagnosis, and thus maintenance and logistics costs.

Interchangeability is another important aspect of design for ease of maintenance of repairable systems. Replaceable components and assemblies must be designed so that no adjustment or recalibration is necessary after replacement. Interface tolerances must be specified to ensure that replacement units are interchangeable.

BIBLIOGRAPHY

1. Patton J D, 'Maintainability and Maintenance Management. Instrument Society of America, Pittsburgh (1980)
2. Goldman A S, Slattery T B, 'Maintainability: A Major Element of Systems Effectiveness'. Wiley, New York (1964)
3. Blanchard B S, Lowery E E, 'Maintainability Principles and Practices'. McGraw-Hill, New York (1969)
4. Cunningham C E, Cox W, 'Applied Maintainability Engineering. Wiley, New York (1972)
5. Nowlan F S, Heap H F, 'Reliability Centred Maintenance'. Dolby Access Press (1979)
6. Jardine A K S, 'Maintenance, Replacement and Reliability'. Pitman (1973)
7. Smith D J, 'Reliability and Maintainability in Perspective'. Macmillan (1981)
8. Arsenault J E, Roberts J A, 'Reliability and Maintainability of Electronic Systems'. Computer Science Press (1980)

19

Solving Reliability Problems

P. D. T. O'CONNOR
British Aerospace
Hertfordshire, UK

INTRODUCTION

Sometimes it is necessary to solve problems in reliability, where the causes of failure are not obvious. This chapter presents an overview of methods which can be used to uncover the causes of such difficult reliability (and production quality) problems. The methods which will be presented are:

1. Pareto analysis.
2. Cause-effect diagrams (Ishikawa diagrams).
3. Analysis of variance (ANOVA).
4. Proportional hazard models.
5. Exploratory data analysis.

PARETO ANALYSIS

Pareto analysis is a very simple form of visual presentation of data which enables one to detect which are the major contributing factors in any situation. A Pareto chart is simply a histogram of the data elements, arranged in order of relative contribution. For example, Figure 1 is a Pareto plot of failure data on a domestic washing machine. It is immediately obvious which are the major contributors to unreliability.

Each contributory factor or component can be further analysed, using the same method. For example, the causes of failure of the programme switch can be further broken down, to show in turn which are the major causes of failure.

Pareto analysis is a very useful and effective technique, partly because of its simplicity. It also enables failure reporting to be planned to be selective, and therefore more economical in finding the causes of the major problems. For example, in the case quoted above, full investigation of the causes of failure might be confined to the top three or four failure modes, including return of the top three components to the manufacturer for detailed examination when they fail.

Common causes of failure must also be evaluated, by looking for features common to different failures plotted. For example, the cause of failure of several different components might be a common assembly or test problem. Therefore the data should be searched for such causes, which when corrected

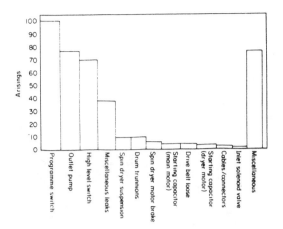

Figure 1.   Pareto Plot of Failure Data

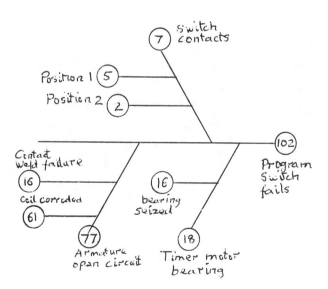

Figure 2.   Cause and Effect Diagram

can reduce the incidence of more than one failure mode.

Pareto analysis is widely used as a basic tool by quality circles teams.

CAUSE AND EFFECT DIAGRAMS

A cause and effect diagram, also called a fishbone or Ishikawa diagram, is a simple way of recording the results of brainstorming sessions for problem-solving. The approach was developed by Professor K. Ishikawa in Japan, as a method for use in quality circles problem-solving.

An example of a cause and effect diagram is shown in figure 2. The approach is that the problem is specified, and possible causes are drawn on the diagram as they are exposed. Each possible cause leads to further contributory causes, until it is decided that the major cause or causes have been identified. These are then actioned. The method is a good way of focussing ideas and recording the results, and its simplicity enables it to be taken up and used with very little training. It is a standard method of the quality circles approach to quality problem-solving.

As possible problem causes are identified, subsidiary diagrams can be constructed for lower-level causes, leading ultimately to identification of the final cause(s).

ANALYSIS OF VARIANCE

Engineers and managers often do not appreciate the effect that variation plays in engineering processes. Whenever individual or combined processes are used to produce an item or to control a production process, or when environmental or other conditions of use can vary, the effects can seriously affect yield, quality and reliability, especially when the process or product is in any way critical. With mass production of multi-process products such as electronics and other high-technology goods, as well as in control of processes such as chemical and metallurgical work, small process or environmental variations can have large effects.

The analysis of variance (ANOVA) method is a very effective way to determine which are the statistically significant sources of variation. It also enables the experimenter to determine which particular combinations of variables, or interactions, might cause large changes in the parameter of interest. This is important, because interactions can have very large effects, and such determinations are often not intuitively apparent.

ANOVA is based on setting up statistical experiments in which the statistical significance of the observed sample variation is tested in relation to the probability that the observed variation could have happened by chance, or is significantly connected with any of the individual process variables. The method is fully described in references 1, 2 and 3.

The method is based upon the fact that when statistically independent sources of variance combine, the combined variance is the sum of squares of the separate variances. (The combined standard deviation is the square root of the sum of the squares of the separate SDs). Thus, for example, if two sources of variance in a process have SDs of 5 and 2, the combined SD would be:

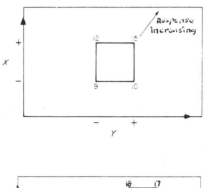

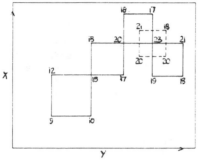

Figure 3.   Response Surface Analysis

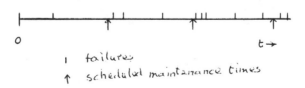

Figure 4.   Exploratory Data Analysis

$$\sqrt{\left(5^2+2^2\right)} = 5.4$$

If we could eliminate the lower source of variation, the resulting SD would be that of the single remaining source, ie. 5.

If, however, we could reduce the larger source by the same amount, to 3, the resulting SD would be:

$$\sqrt{\left(3^2+2^2\right)} = 3.6$$

Thus, by reducing the larger source of variance instead of the smaller, a much larger decrease in total variance is achieved.

ANOVA can be used for analysing complex situations involving several variables. A typical modern application is the optimisation of wave soldering in electronics manufacture, in which the variables are solder temperature, wave height, speed through the wave, solder mix, and features of the assemblies being soldered.

A related technique is <u>response surface analysis</u>, a method for optimising processes once the key variables have been isolated. Experiments are conducted in which these variables are set at different levels, and the output is measured. After each set of experiments, with the variables each set to a high and a low value, the experiment is then repeated with the variables reset to follow the line of best overall response. The process is repeated successively until the best output is achieved. Figure 3 shows the method graphically. Response surface analysis is also called <u>evolutionary operation</u> or <u>hill climbing</u>.

The mathematical methods used for ANOVA are rather tedious. However, computer programs are readily available for analysing the experimental results.

PROPORTIONAL HAZARDS ANALYSIS

Proportional hazards analysis (PHA) is a mathematical method for determining the major contributors to an observed effect, for example the hazard rate of an item. The method assumes that the overall hazard function is the product of a base-line hazard function and an exponential term which includes the effects of the explanatory variables, $z_1, \ldots z_k$. That is,

$$h(t; z_1, z_2, \ldots z_k) = h_0(t)\exp(\beta_1 z_1 + \ldots \beta_k z_k)$$

The values of β_1, β_2, ... are the values of the parameters which define the effects of the explanatory variables.

The baseline hazard function can be assumed to be Weibull distributed, or a non-parametric approach may be used.

The method can be used to identify individual significant contributing factors, as well as interactions. It is described in references 4 and 5. As with ANOVA and other statistical techniques, computer programs are available to enable the analysis to be performed quickly and accurately.

EXPLORATORY DATA ANALYSIS

Exploratory data analysis (EDA) is a term used here to cover a range of techniques whereby reliability (or other) data are assessed for trends or patterns, using relatively simple techniques, a common feature of which is visual interpretation of plotted results.

Point process analysis, as described in Chapter , provides the basis for one such technique. For example, Figure 4 shows times at which failures might have occurred on an equipment subject to regular scheduled maintenance. The scheduled maintenance interval is shown on the figure. The data show that there is a bunching of failure events after each scheduled overhaul, indicating that the overhaul work might actually be introducing unreliability.

This is a very simple example. In other cases different failure modes might be separately plotted, or the data for a number of systems might be plotted together, with the axis normalized to superimpose overhaul intervals, seasons of the year, or other feature that might be considered significant. For example, a fairly common feature of repairable system reliability data is bunching of failure events after repairs, when repair actions (including diagnosis) are not always effective the first time.

Further EDA techniques which can be used to test for patterns (or "structure") are cumulative sum plots (CUSUMS) and probability plots (as described in Chapter). Reference 6 provides a good description of EDA methods.

CONCLUSIONS

This brief chapter has outlined methods available for solving reliability problems. In fact the methods do not by themselves solve problems, but they can be used to support engineering judgement in deriving the likely causes of problems and for optimising the corrective or control actions. It must be stressed that they cannot and must not be used without the support of the appropriate engineering and scientific knowledge. Thus, the methods do not provide explanations; they only provide clues. Nevertheless, these clues can be very strong, and can lead to the solution of otherwise intractable problems. The ready availability of computer programs for the more complex methods make their application effective and economical.

REFERENCES

1. O´Connor, P.D.T., Practical Reliability Engineering, (2nd edition), J. Wiley 1985.

2. Box, G.E., Hunter, W.G., and Hunter, J.S.,Statistics for Experimenters, Wiley, 1978.

3. Lipson, C., and Sheth, N.J., Statistical Design and Analysis of Engineering Experiments, McGraw Hill, 1973.

4. Dale, C., Application of Proportional Hazards Models in the Reliability Field, Procs. 4th UK National Reliability Conference, 1983.

5. Bendell, A., Walley, M., Wightman, D.W., and Wood, L.M., Proportional Hazards Modelling in Reliability Analysis- an Application to Brake Discs on High Speed Trains, Quality and Reliability Engineering International Vol. 2 No. 1 1986.

6. Bendell, A., and Walls, L.A., Exploring Reliability Data, Quality and Reliability Engineering International, Vol. 1 No. 1 1985.

20

■

Management of Reliability

P. D. T. O'CONNOR
British Aerospace
Hertfordshire, UK

INTRODUCTION

Since the quality and reliability of a product are affected by so many
activities and people, from the specification, through design, test,
production and support, high reliability can be achieved only by an
integrated, all-embracing effort that ensures that there are no weak links in
the entire development and production process. Such a comprehensive approach
can be applied successfully only from the top of the organization concerned.
Therefore, quality and reliability must be essential elements in the corporate
strategy and top-level direction. This in turn implies that the corporate
head understands the causes and effects of defects in production and failures
in service, and the methods to control them.

This chapter discusses some aspects of the management of reliability
programmes. The references cover further aspects of the topic.

COSTS AND PRODUCTIVITY

Quality and reliability, like any other product feature, must be measured
ultimately in financial terms, since this is the ultimate measure of business
success. However, financial success is not measured only in short-term
performance, but in long-term considerations such as survival of the
enterprise, market share, future product and marketing strategies, and
security and oquality of life for the employees. Quality and reliability have
a major effect on these long-term business objectives, and only long-term
strategic action can ensure that quality and reliability match the
requirements of modern markets.

As explained in Chapter 5, a design that is created to be reliable, by an
integrated engineering team motivated and equipped to prevent design
weaknesses and to detect and eliminate them as early as practicable in the
development programme, will be less costly to bring to the production stage.
In other words, design and development for reliability improves engineering
productivity. In Japan, the concept of design to maximise reliability, taking
full account of the production process capabilities, is called "off-line
quality control".

Production quality is a major determinant of productivity, and thus of
profitability. Managements usually claim to understand productivity and
profitably, yet they often do not appreciate how seriously these are affected
by quality.

Every defect on a production line creates the following activities or other costs:

1. Repair or scrapping of the defective item.

2. Documentation of the repair or scrap. This includes accounting for the materials and components, instructions on action to be taken, ordering of replacement parts, analysis of defect data, etc.

3. Retest and re-inspection, and associated documentation.

4. Extra inventory (work-in-progress, or working capital), represented by the materials, components and value added to the defective item.

5. Floor space, storage space, and all the other burdens associated with having the defective item on the premises.

6. Supervisory and management time and effort in managing, recording, discussing, explaining and trying to correct the defect and trying to prevent recurrence.

7. Possible delays in satisfying orders.

8. Possible failures in service, since repaired products are often less reliable than those that have not needed repair, and defects that can occur in production can often lead to failure in service.

All of these costs can obviously add greatly to the costs of production, since all the resources described above are not available for actual production. Furthermore, since the activities involved are never routine, they involve a disproportionate amount of effort and resources. It is not possible to plan effectively and to budget accurately for failures, since it is not possible to know or to forecast what they may be and how serious might be the effects. Consequently managements must live by "fire brigade" actions, and console themselves that they are taking appropriate action by holding meetings, setting targets for improvement, exhorting staff to do better, and blaming workers and suppliers. It is not uncommon for the costs of living with defects to add 20% to the costs of production. This represents that amount of loss of profit or the appropriate loss of sales which could be obtained by reducing price.

Of course it is not possible to obtain totally defect-free production of complex items. Nevertheless, defect-free production must be the objective. Anything less implies that there is an "acceptable" quality level of less than 100%, in other words, imperfection is tolerable. This attitude leads to the "fire brigade" mentality described above, resulting in a situation which is not really under control.

High quality of the design and of production both increase productivity, by reducing waste of resources, materials and management time. They also lead to direct cost reductions, both during development and production. Therefore the old concept of an "optimum" level of quality and reliability, less than zero failures, is misleading and defeatist. The modern approach recognises that the only optimum is continued improvement towards perfection, as shown in Chapter 1 (Figure 1).

PROCEDURES AND METHODS

Management response to out-of-control situations is often to introduce
procedures. Procedures might cover such aspects as defect recording,
reporting and analysis, calibration of instruments and gauges, control of
bought-in material, etc. When a problem arises, either the procedures or the
people who are supposed to work to them are blamed. Procedures might be
changed or people reprimanded or otherwise dealt with. Management takes no
blame upon itself. Yet management must take responsibility, since only
management can provide the methods and the motivation to create excellent
products.

Methods are much more important than procedures. Methods include the machines
and other equipment to do the job, as well as techniques such as statistical
process control (SPC). Only management can supply these, by purchase and by
training. Management people can usually understand the need for machines to
improve productivity. However, they have great difficulty, particularly in
the west, in understanding the essential part that process control methods
should play. The Japanese have somehow grasped this more readily, ironically
mainly from the teaching of Americans, notably Deming and Juran.

PROCESS CONTROL

Process control involves the control of variation. Variation can only be
controlled by applying statistical methods. Statistics can be used to provide
assurance that specified maximum proportions defective are not passed or
accepted ("acceptable quality levels"), or in other risk-reduction
applications such as reliability demonstration testing. However, statistical
methods can be used in much more positive ways, to control variation in such a
way that compliance is total, and no defects are produced. It is this power
of simple statistical methods that the Japanese have applied with such great
effect. However, to apply these methods requires management knowledge of how
they work and the will to apply them across all processes which can vary, and
which can therefore lead to defects or subsequent failure if not controlled.
The Japanese call this "on-line quality control".

The concepts of off-line and on-line quality control, in which the design is
optimised in relation to variations in the production processes and variations
in the environment, and then the production processes are controlled to keep
their variations within the tolerances assumed in the design, is based upon
the use of statistical methods to produce highly reliable, robust products
with extremely high productivity and quality. The basic methods are simple,
as has been shown in earlier chapters. What is notable is the comprehensive
way in which they have been applied by the most advanced companies, under the
"total quality control" or "company-wide quality control" approach. The
challenge to management is to ensure that the methods are applied to all
design stages, all testing and all production processes. For managers trained
in the traditional ways this is indeed a challenge, and it is not surprising
that many companies and other organizations have found it difficult to adopt
the new methods.

REQUIREMENTS AND ASSURANCE

It is essential that the quality and reliability requirements are stated in
ways that are meaningful and therefore realisable.

Reliability requirements can be expressed as availability or mean time between
failures, but the factors that influence these parameters are very much

affected by human actions and interpretation, as is explained in Chapter 15 on reliability prediction. For a manufacturer, a more useful parameter might be the cost of warranty, since this is a parameter which is easily monitored and which being a financial index is directly comparable with other costs. Other parameters could be the warranty return rate for repair or replacement.

The problem is greater for the user of the product. If he specifies the product and pays for all or part of the development he must also specify the reliability. The standard approach for many years has been to specify an MTBF, to be demonstrated during development and production, and sometimes also in service. We have covered the problems of reliability demonstration testing in Chapter 14, and the methods described in documents such as US MIL-HDBK-781D must be considered inappropriate. However, reliability demonstration in service also has major drawbacks:

1. It gathers information after the event. If the achieved MTBF, measured over a sufficiently (and therefore usually inordinately) long time is significantly below the specified level, the damage has been done in terms of cost and other inconvenience, and it is usually very difficult or impracticable to make any significant improvement to equipment in service.

2. The data collected is nearly always subject to even worse problems of interpretation, effects of user and maintainer actions, incorrect diagnosis, etc., than is such data gathered in factory testing. This more than cancels the gain in realism due to the fact that the product data is generated in a real as opposed to an artificial environment.

It is therefore very difficult for a user to monitor effectively and accurately the reliability performance of equipment in service. He will be very aware in a qualitative way, but this is not sufficient for ensuring that corrective action will be taken by the supplier, and, as stated above, it is usually too late anyhow. How then can the user assure himself that he will receive reliable equipment, and how does he state his requirements in order to achieve this?

The user´s requirement must be stated in such a way that the supplier will both understand and be motivated to act upon it. It is therefore necessary to avoid the semantic and other traps described above, and relate the reliability in service to a perceived financial objective. The only way to achieve this effectively is to transfer the cost of failures, or a part of the cost, to the supplier, so that he shares the risk and life cycle costs involved. In this way the supplier is motivated at the start of the project, when he can plan and possibly design to minimise his, and therefore the user´s, costs. This cannot be achieved by specifying an MTBF, but, in keeping with the principle of design and manufacture for no failures, by specifying that the supplier will correct all failures at his expense, and in a specified time. In other words, the supplier provides a warranty. There is then no delay while reliability statistics are collected and analysed, and no argument about achievement. No reliability level below excellence becomes "acceptable".

Warranty contracting is, of course, the norm in very many purchase situations, particularly for domestic and much industrial equipment. Extending warranties to products such as major plant and defence equipment, which must operate and be maintained for long periods, often under conditions not anticipated by the supplier, obviously can present difficulties. However, many of these can be overcome to a sufficient extent to enable the principles and the payoffs of warranty contracting to be realised. Warranty contracting is discussed in more detail later.

Requirements for production quality should follow the same principles. The user should demand nothing less than perfection, and therefore concepts such as "acceptable quality level" should be rejected in favour of the supplier replacing or repairing all defective products.

For the manufacturer, stating internal production quality requirements must also be based upon the same principles. There is no production quality problem which cannot be prevented or corrected, so the requirement of each production process must be that it is controlled within the specified process limits, which in turn ensure that all production is to specification. All deviations must be identified and brought back under control immediately. Note that this is not the same philosophy as the "zero defects" approach once popular particularly in the USA. ZD is based upon target setting, exhortation, presentations and propaganda. It assumes that motivation is enough to create consistent perfection. However, ZD provides no methods. Therefore it can lead to frustration when the objectives are not met, and the reasons why are not understood or are "someone else's fault". Its success depends upon continuous publicity and management involvement, and this can be hard to sustain at the initial level of enthusiasm and effectiveness. The modern approach to production quality control is based primarily on providing the means whereby processes are kept in control, through methods, training, and motivation. By providing methods and training a more secure base for quality control is created, not dependent on high-level exhortation. The people involved at the working level become the masters of the processes, since they understand the sources of variation and how to control them. They are much better equipped to detect and correct problems, or to advise on the best corrective action. Finally, the quality control system sustains itself. All of these features are part of the quality circles movement, as described in Chapter 17.

ETHICS IN RELIABILITY

Since reliability cannot be measured accurately, and since very large amounts of money might be involved in design, testing, provision of spares and maintenance, there is inevitably scope for divided interests to operate to the detriment of reliability. For example, if a complex system is purchased, with no warranty but with a reliability "target", the purchasing agency might be reluctant to admit to a shortfall if it reflects on their management of the project, and the supplier might be pleased to provide the additional spares and other support which must be paid for. Alternatively, if the purchaser expresses dissatisfaction and attempts to have the supplier take corrective action, there is inevitably argument about responsibility for failures and even about which events should be counted as failures. Formal reliability demonstrations can also suffer from this type of vested-interest based argument, regardless of how tightly specifications and definitions are laid down.

Human nature being the way it is, the only solution to the problem is to ensure as far as is practicable that both sides are motivated towards the same objective of high reliability. This can be achieved by proper attention to how reliability is covered contractually.

CONTRACTING FOR RELIABILITY

The simplest and most common form of reliability contract is the straightforward warranty. In competitive situations, for example in consumer goods and most industrial equipment, warranties provide very powerful motivation for high reliability, and the customer need only use his freedom of choice and take advantage of his warranty when necessary. When purchasing agencies buy products which are normally sold with a warranty, for example state agencies buying vehicles or office equipment, they can usually obtain the best protection for their interests by using the standard commercial warranty offered to any other customer, and they need only set up the appropriate organization to exploit it.

For higher risk projects, such as defence equipment and large plant installations, careful attention must be paid to how reliability is covered in the contracts. Such systems usually involve multiple contracts covering a range of aspects, from high-risk development to maintenance and support, and can also include low-risk commercial items. The purchaser needs to assess which aspects will present the greatest risks and life cycle cost implications, and then plan his contract and negotiating strategy accordingly. In view of the great uncertainty inherent in reliability prediction and demonstration, this is not a straightforward task, but the implications of oversights or undue optimism at this stage can be very large indeed.

Some of the main guidelines to be followed are described below.

Lowest cost contractor selection is probably the greatest contributor to future trouble and cost. Unfortunately many official organizations tie themselves to this outmoded purchasing concept, despite the overwhelming evidence that it so often leads to much greater total cost of ownership. Contractors should be selected on the basis of reputation, quality of the total submission, and committment to future support. This is increasingly the method used by domestic and industrial consumers, and it has led to the great increases in quality and reliability in these markets. Unfortunately, contractor selection on the lowest first cost basis can be a simple way of convincing taxpayers that the selection procedure is fair, so other methods should be introduced to ensure that only the appropriate factors are considered. For example, independent contract review boards can be set up, to review the reasons for selection before the contract is let. Such review boards should be made up of competent people, able to understand the technical and financial details. To ensure integrity, they could be called together from a list of competent people after the bids have been submitted, and sit as a jury.

It is often possible to include life cycle support as part of the procurement contract, and selection can then be based, at least in part, on the lowest life cycle cost as opposed to lowest first cost. LCC contracts are becoming more popular with purchasing agencies in appropriate circumstances. The contract should cover provision of support and spares for a defined period. In this way the supplier will bear at least part of the costs of unreliability, so that motivation is ensured.

A particular example of a type of LCC contract is reliability improvement warranty (RIW), a form of contract used for both military and civil equipment, particularly in the USA. For example, the avionics of the F16 fighter aircraft in service in the US and several European air forces are covered by RIW contracts. In the RIW system, the supplier provides all necessary repairs to sub-assemblies, and provides spare serviceable units "off-the-shelf", for a period of typically four years for a fixed up-front payment. In this way the user is spared the cost and difficulty of setting up repair depot facilities, and of estimating spares requiremens and having to provide these. The supplier is obviously highly motivated to provide a reliable system. The cost to the user is fixed whatever the reliability, and budgetting is therefore easier.

300

LCC and RIW contracts must be negotiated and applied with skill and flexibility. The particular system and applications need to be taken into account. For example, it must be possible for the supplier to perceive the risk involved, and he might reasonably expect to have some control over the way the system is used and maintained, and to have access to it. Often the contracts can usefully include training aspects, to ensure that the user is properly equipped to maintain and operate the system.

In some cases an incentive payment can be made for reliability achievement. For example, such payments have been made in the past to suppliers of communication satellites for successful operation in orbit for a given period. Such incentives have the advantages that they are simple to negotiate and administer, since the customer can simply state his intention, and he is in control of the data on which the payment will be based. The supplier has the motivation that he will be paid for performance, not penalised for failure, and therefore he is not therefore in a position to argue about the result.

It is most important that reliability contracts do not become burdened with unverifiable quantitative reliability requirements, for the reasons described above. Contracts must emphasise practical, motivational aspects, otherwise their implementation can result in fruitless arguments about the extent to which quantitative requirements have been met. The use of statistical confidence limits should be avoided, since they can lead to further difficulty of interpretation.

STANDARDS AND AUDIT

For some projects, it might be appropriate to impose a reliability standard, such as US MIL STD 785 or British Standard 5760. These standards, when accepted as part of a contract, require the supplier to operate a reliability programme which includes reliability prediction, design reliability analysis, reliability testing, and failure reporting and analysis. He is also required to operate a management system to co-ordinate and control these activities, and to report on them. These standards can be used to ensure that a basic minimum awareness exists, covering the activities necessary. However, the standards do not by themselves ensure that the product will be reliable. In this way they are similar to standards for quality assurance. The standards ensure that systems and procedures exist, but they do not guarantee that they are effectively operated. They are often out of date in relation to new approaches, such as quality circles and robust design. Therefore they can provide a false sense of security, and their application can be used as an excuse for increased cost. They are not a substitute for effective management and the application of effective methods.

Suppliers can be audited to ensure that they comply with the requirements of a reliability programme, and the main national quality and reliability standards include supplier audit as one of the recommended activities. Audits can be very useful or they can be ineffective and expensive, depending on the approach taken. Audits should concentrate on the methods used, not merely on compliance with procedures. They should also cover competence and motivation. Therefore the staff conducting the audits must be competent to do so. The audit must be related to the project and the nature of the contract. For example, there is little to be gained by carrying out a detailed audit of a supplier's procedures if the project is to supply a standard item with a good warranty. On the other hand, it might be appropriate to audit the methods of a supplier of a new type of product if the supplier has not been able to show evidence of past good performance and if the contractual motivation for reliability is not very strong.

CONCLUSIONS

This chapter has discussed only a few aspects of reliability management. More detailed information can be found in the reference. The overriding requirements of good management of reliability are:

1. An integrated approach to all engineering, production and support tasks.

2. Clear, unambiguous statement of the reliability objectives, without undue emphasis on the precise quantitative aspects.

3. A contractual framework that provides motivation for reliability.

High quality and reliability are the natural results of competition and freedom, in which the consumer is free to choose and to base his choice on excellence of design and manufacture. In turn, recent developments have shown clearly that excellent design and production are not more expensive in fulfilling a given need. Therefore, the choice of a supplier for a product should take into account the extent to which the potential supplier has embraced the modern approaches to quality and reliability, and the support he will provide. Experience generally shows that the products of centralised State organizations do not match the reliability of products from Western and democratic far-Eastern companies. Dealing with a company rather than with a government is nearly always safer and results in more reliable products and service. A company has no ideology, except to continue in business. Therefore companies are inherently motivated to supply reliable products at reasonable prices, and to support their customers. This is a major advantage enjoyed by the democracies, and consumers should exploit this to their own benefit.

REFERENCES

1. O'Connor P.D.T., Practical Reliability Engineering, John Wiley and Sons Ltd. (2nd edition, 1985).

2. Deming J.R., Quality, Productivity and Competitive Position, MIT University Press, 1983.

Index